They said computation was clean.
They said it was weightless.
They lied.

Artificial intelligence was supposed to be the future. It may be the last thing the grid can afford.

Keep Computing

How Light Solves Computing's Impossible Problem

Derek W. Bailey

TRUE PHOTONIC, INC.

Santa Rosa, California

Published by True Photonic, Inc. 1260 N. Dutton Street Santa Rosa, CA 95401
www.truephotonic.com

ISBN: 978-0-692-98288-4
Printed in the United States of America
978-0-692-98288-4

Dedication

For my daughters, Alexandra and Maxine, who have paid all the tolls on my journey road.

For my mother Ozie-Mae, my late grandmother: Josephine, and my brothers Orrin and Walter, my belief in myself derives from you.

For Pattie Wexler - best friend- who keeps me balanced.

And for everyone who was told it couldn't be done—**go build it anyway.**

Acknowledgments

A book like this does not happen alone. Neither does a company.

Dr. Gary Poovey trusted us with his life's work. That trust made everything else possible.

Del Wolverton has been my partner for twenty-seven years. Everything in these pages that works—the technology, the architecture, the vision—carries his fingerprints. Brother, thank you.

Roy Geddie, CEO of WolvertonBailey, has been a constant force moving this project forward. His steady hand keeps us grounded when the vision gets ahead of the execution.

Mark Newland leads our project management team with the discipline and precision the work demands. He was one of the first engineers at SanDisk; we are fortunate he chose to build something new with us.

Michael Miles, Steven Branch, Taylor Paul, Alex Hicks, Nathaniel Anderson, and Douglas Eze have worked across both Derek Automotive and True Photonic, supporting me at every step.

Robert Switzer and the team at Astera Energy brought us the substrate of the future. The graphene-boron nitride-sapphire architecture that enables the Poovey Stack's thermal performance exists because of their work.

Former Congressman William Jefferson provided guidance and opened doors as we work to bring True Photonic to New Orleans. His belief in what photonic computing can mean for the Gulf South has been invaluable.

Carlos Austin has been a steady advisor—the kind who tells you what you need to hear, not what you want to hear. "Your suggestions made this a better book."

Mike Rodriguez made the introduction that started everything—a handshake at a small church in Mt. Shasta that changed the trajectory of three lives and, we hope, an industry.

Searcie Cassidine of Cassidine Consulting brought rigor to our IP valuation and insisted I write this book. She was right on both counts.

To the engineers, attorneys, advisors, and staff across True Photonic, WolvertonBailey, BizBrolly, Modern Fortis, and Derek Automotive who do the work that makes the vision real—thank you. Your names may not all appear on these pages, but your contributions appear on every one.

And to my family, who have lived with "impossible" projects for decades: I know the cost.
I do not forget it.

CONTENTS

Dedication 3

Acknowledgments 6

Prologue: This Is Not My First Impossible System 9

PART I: THE IMPOSSIBLE PROBLEM

Chapter 1: The Cost of a Question 19

Chapter 2: The End of Moore's Law 27

Chapter 3: The Weight of the Cloud 35

Chapter 4: The Vulnerable Stack 43

Chapter 5: The Backlash 51

PART II: THE NATURE OF LIGHT

Chapter 6: What Light Can Do 67

Chapter 7: The History of Photonic Computing 79

Chapter 8: The Breakthrough 85

PART III: THE ARCHITECTURE

Chapter 9: The Poovey Stack 97

Chapter 10: Logic-in-Light 111

Chapter 11: Building the Machine 125

PART IV: THE FUTURE

Chapter 12: Clean Compute 141

Chapter 13: The Next Stack 153

Chapter 14: The Light Ahead 169

Afterword: Infrastructure That Shares 181

For forty years, photonic computing was a solution in search of a problem. The problem has arrived.

Prologue

This Is Not My First Impossible System

Dr. Gary Poovey spent twenty-five years inside semiconductor fabs, watching photonic computing fail.

He worked at Lam Research, at Bullen Semiconductor, at companies whose names appear on the equipment that manufactures every chip in every device you own. He understood plasma physics at the doctoral level. He understood the practical realities of wafer fabrication at the production level. And he understood, with painful clarity, why every attempt to make light compute had fallen short.

The switches were too slow. The materials were too exotic. The voltages were too high. One promising approach required 6,000 volts to invert a signal—enough to kill a person, utterly impractical for commercial deployment. Another worked only at cryogenic temperatures. Another required materials so expensive and difficult to fabricate that no business model could justify the investment.

Gary kept working. He designed seventeen different light switch structures and had them fabricated for testing. Most failed. Two switched light at least twenty times faster than a transistor—forty picoseconds, fast enough to matter. He demonstrated that piezoelectric materials respond to the electric field of light itself, not just to applied voltage. He filed patents: eight granted, more pending. He built the foundation for something that might actually work.

By 2022, Gary had the physics. **What he did not have was a path to market.**

He was approaching eighty. The funding sources that back early-stage technology do not typically write checks to octogenarians, regardless of how sound the science. His research had consumed years and millions of dollars. The Technion report sat in his files—independent validation from one of the world's leading materials science institutions, documenting switching speeds of 150 to 200 femtoseconds, three to four times faster than anything in the published scientific literature. It was proof that the physics worked. It was also, in a sense, an ending. A research program with nowhere left to go.

Then Mike Rodriguez made an introduction at a small church in Mt. Shasta, California. This was 2022.

Mike, a friend and WolvertonBailey shareholder, introduced Gary to Del Wolverton—Chief Science Officer at WolvertonBailey, a Navy Senior Chief with decades of experience in engineering, radar, missile systems, and photonics. Del had founded Space Microwave and built fiber optic switching technology that shipped in commercial telecommunications systems. He understood what it took to bring optical technology from laboratory to market.

Del did not know Gary was a physicist. Gary did not know that Del had received a DARPA grant for work that made light do impossible things. They were just men in a congregation, exchanging pleasantries.

Mike mentioned that Dr. Poovey was looking for funding for some research. Del listened to Gary's quick story about his work. It registered immediately. He invited Dr. Poovey to our lab.

A few weeks later, Gary came to Santa Rosa. He brought data—not theories, not projections, but the validation report from Technion University in Israel. His team had tested optical switches built from a material he had developed, one that responds to the magnetic field of light. The measured switching speed was 150 to 200 femtoseconds. The fastest optical switching speed ever published in scientific literature was around 480 femtoseconds. Gary's material was three to four times faster than anything that existed.

Del called me. As he always does when he sees something I should see. He described the encounter, the report, the background. Before he could finish, I asked: "Couldn't we make a transistor with that?"

"Exactly," Del said. "But say nothing to Dr. Poovey right now. I don't think he knows what he really has. I'm going to recreate the Technion test in our lab. If I get the same result, we gotta move."

"How far along is he?"

"He has a first-generation optical switch. But no logic. His research team was strong, but there were no computer guys on it. And the material composition is not optimal." Del paused. "I can help him fix all of it."

"Then let's get this funded," I said. "At Derek speed."

Derek speed is an inside joke at the company. It means now.

I do not need the whole picture. I trust Del's technical expertise, and I look to invest while the idea is still on the napkin. That is what separates the daring from the prudent. Most investors wait for proof. They want validation, traction, a clear path to returns. They want someone else to have taken the risk of being wrong.

By the time they see opportunity, the opportunity has been priced accordingly. I was already partnered with the critical technical guy. Del and I had spent twenty-five years building things that were not supposed to work:

CopperSnake—video over copper wire when the industry said copper was dead.

Popa Media—digital billboards installed on decommissioned payphone stands across urban centers, offering free phone calls through voice-over-IP. This was when people were still paying for minutes on their cell phone plans. They could not understand how we could give calls away for free. We understood. We were just early.

The Counterpoise engine—a combustion system that defied what the textbooks said was possible, now patented, 83 percent more efficient than the diesel cycle. Each project demanded we solve problems no one else had solved because no one else had attempted them. When I looked at Gary's Technion report, I did not see a physics result. I saw an incomplete picture that we knew how to complete.

Gary Poovey had a switching material. He did not have a transistor. He did not have logic gates. He did not have an architecture, a manufacturing strategy, or a path to market. He had the first piece of something that could change computing—and no way to build the rest.

We had the way.

I saw immediately that this technology was more important than anything else I was working on—including the automotive work at Derek Automotive that had consumed years of effort and investment. The self-charging hybrid vehicle I had been developing required computation at speeds and efficiencies that electronic systems could not deliver. The processors would be obsolete before the car reached production. The foundation was wrong.

With a photonic computer, that vehicle could become fully autonomous. Real autonomy—not the assisted driving that companies market as self-driving—requires computation that electronic systems cannot provide at acceptable power levels. I had been trying to solve the automotive problem without first solving the computation problem. Gary's breakthrough meant I could solve the foundation first. The automotive work would still be there, and it would be better for having waited.

Del would provide the engineering expertise through WolvertonBailey by creating a special photonic's department. I would build the narrative and the business structure—which for an idea this ambitious, in an industry as entrenched as semiconductors, would have to be as novel as the technology itself.

That is what True Photonic, Inc. is.

That is what True Photonic Ventures will become: a venture arm that seeds startups built on our photonic platform, companies that will locate their operations in our SkyLight Clean Compute Centers. Not a venture fund that waits for founders to figure everything out, but a venture partner that brings the capabilities to complete pictures others cannot. We will not just write checks. We will finish ideas and architectures.

We made Gary an offer. He did not accept immediately. He went home and prayed with his family. Gary Poovey is a deeply religious man, and he was being asked to entrust his life's work to people he had met at church a few weeks earlier.

The decision deserved more than a handshake.

When he called back, the answer was yes—**with a condition.** He wanted assurance that we would be a different kind of company. An ethical company. In his words: "Christian businessmen."

The trust that required still humbles me. I made a promise to Gary's wife: I would make their family name famous around the world. And so I named the light-actuated light-switch—the device at the heart of everything that follows—the **Poovey Switch.**

It is a promise I intend to keep.

Dr. Gary Poovey merged his technology into the company as Chief Science Officer. He leads the research and development that translates the fundamental switching mechanism into manufacturable devices. The intellectual property he developed—new compositions and gate architectures that address the limitations that defeated earlier photonic approaches—represents fundamental innovations protected by the patent portfolio described throughout this book.

Del assembled the engineering team. First was Mark Newland. Mark was one of the early engineers at SanDisk, part of the team that built the flash memory industry from laboratory curiosity to global infrastructure. His career had traced the arc of modern computing—a bachelor's degree in applied mathematics and computer science from California Polytechnic, an MBA in computer information systems from National University, decades of software development in C++, C#, and Linux environments. He had led cross-platform application engineering efforts in both industry and academia.

Mark understood what it takes to move technology from prototype to production. He had done it before, at scale, at a company whose products now lived in billions of devices worldwide. When Del explained what the Poovey Switch could become, Mark saw what we all saw: not merely a faster switch, but the foundation of a new computational paradigm.

A physicist, an engineer, and a dreamweaver walked into a lab. It sounds like the setup to a joke. **The punchline is that it worked.**

On December 12, 2024, the United States Patent and Trademark Office published our application for a purely optical NAND gate—**the universal building block from which all Boolean operations can be constructed**. Gary, Del, and a team working together wrote that patent. Del's architecture informed it. The publication number is US 2024/0411203 A1; anyone can search it and verify the claims. The optical NAND represents what decades of photonic computing research sought but failed to achieve: a manufacturable, all-optical gate capable of implementing any Boolean function. Gary's physics made it possible. Our team made it real.

• • •

I have not received a paycheck—one that was not 100 percent commission—since I was nineteen years old. I am no longer a spring chicken.
That is not a boast. It is context.

When I tell you that impossible tasks and setbacks do not frighten me, I am not speaking from theory. I am speaking from more than four decades of operating without a net, without institutional backing, without the assumption that effort would be rewarded simply because it was applied.

Hospitality management came first. Then real estate. Then insurance. I became a registered investment representative, then left to sell home improvements.
I owned a restaurant. I ran food trucks. I built businesses in Asia and Latin America, flirted with business in Russia—export and marketing operations that taught me how systems work differently in different places and how they fail for the same reasons everywhere.

The formula I learned: **Have a vision. Build a narrative. Go sell that shit.**

The breakthrough came in temporary staffing—a business that scaled, generated real capital, and still operates today. Most people, at that point, would have consolidated. Stayed in their lane. Protected what they had built. The family has always hated the way I put us at risk. But I have never been good at staying in lanes.

As a kid, when adults asked what I wanted to be when I grew up, I said I wanted to own a company like GM. I am not sure I even knew GM made cars. I just remembered commercials that said something like "What's good for GM is good for America."

That sounded right to me. I want a company like that someday, I would tell them. People would laugh. **This kid.**

The staffing profits went into technology. Then into vehicles. Now into the problem that underlies everything: **computation itself.**

CopperSnake was not supposed to work. The industry consensus was that copper could not handle video. Too much impedance. Bandwidth too narrow. The physics did not support it. We built it anyway. Video transmission up to a mile over copper wire. That was my first lesson in the difference between "impossible" and "not yet attempted by the right team."

The Counterpoise engine was not supposed to work either. Del came to me with a design that defied what the textbooks said was possible. I was on a 7:30 a.m.

flight from Philadelphia. I knew nothing about engines. But I knew Del. WolvertonBailey, Inc. was formed to develop the Counterpoise engine—a generator system that proved to be 83 percent more efficient than the diesel cycle. It completes its rotation in 360 degrees instead of 720. It is a third the size for equivalent power. It uses half the fuel. Now patented.

True Photonic is the culmination of everything I have learned about building companies around impossible technologies. The physics is sound. The engineering is underway. The team is assembled. The patents are filed. The facilities are being prepared.

The title of this book contains both a **description** and a **command**.

Keep Computing.

The world needs computation to continue advancing. The current infrastructure cannot sustain that advancement. Something must change, or progress stops. Gary Poovey spent twenty-five years understanding why photonic computing failed. Then he invented something that works. Our engineering team recognized it. I'm working to fund it. Mark Newland is guiding the building of it. Robert Switzer at Astera Energy is developing the substrate that makes it inexpensive to manufacture and provides an incredible building platform. Investors, friends with professional, legal, lobbying, and political backgrounds believed in the infrastructure vision early enough to matter.

The pieces are assembled. The picture is nearly complete. What follows is the story of why this matters—**and what we intend to do about it.**

— Derek W. Bailey
Atlanta, Georgia
January 2026

The data center is the factory of the twenty-first century. And like every factory before it, it extracts.

PART I

THE IMPOSSIBLE PROBLEM

The water did not come back.

Light does not heat
what it touches.

CHAPTER 1

The Cost of a Question

You asked a question.

Perhaps it was idle curiosity—the capital of Burkina Faso, the lyrics to a half-remembered song. Perhaps it was consequential: a medical symptom that needed interpretation, a contract clause that required explanation, a line of code that refused to compile. Whatever the question, you typed it into a box on your screen, pressed enter, and within seconds received an answer that would have taken your parents a trip to the library, and your grandparents a consultation with an expert they may not have been able to afford.

The answer appeared as if by magic. Clean text on a white screen. No visible machinery. ***No evident cost.***

But there was a cost. There is always a cost.

In the fraction of a second between your question and its answer, a cascade of physical events unfolded across thousands of miles. Your query traveled through fiber-optic cables to a data center—perhaps in Virginia, perhaps in Iowa, perhaps on three continents simultaneously. There, it was routed to a cluster of servers stacked in racks that stretch toward forty-foot ceilings, each server packed with processors running so hot they would melt without constant cooling. Those processors performed billions of calculations, drawing electricity from a grid that may have burned coal or natural gas to generate it.

The heat they produced was absorbed by water—thousands of gallons of it—which carried the thermal energy away from the machines before evaporating into the atmosphere or cycling back through cooling towers the size of apartment buildings.

Your answer materialized. You read it, nodded, and moved on with your day.

The water did not come back.

The infrastructure that answers our questions is, by design, invisible. This is a triumph of engineering and a catastrophe of perception. We have built the most energy-intensive industrial system in human history and hidden it so completely that most of its users do not know it exists. The system is called "the cloud," a word chosen for its weightlessness, its ethereal connotation of something floating benignly overhead. The metaphor is a lie. The cloud is not floating. The cloud is a constellation of windowless buildings, many of them larger than aircraft hangars, consuming more electricity than small cities.
The cloud requires its own power plants. *The cloud drinks rivers.*

A modern hyperscale data center—the kind operated by Google, Microsoft, Amazon, and Meta—covers between 500,000 and two million square feet. Inside, tens of thousands of servers operate continuously, processing the queries, storing the data, and running the applications that constitute our digital lives. A single facility can draw 100 megawatts of power or more. For context, that is enough electricity to power approximately 80,000 homes. It is the output of a small nuclear reactor, or a medium-sized natural gas plant, dedicated entirely to computation. And computation generates heat. The processors that answer your questions operate by pushing electrons through silicon pathways so small they are measured in billionths of a meter. This is violent work at the atomic scale. Resistance generates heat. Switching generates heat. The very act of thinking—if we may call it that—generates heat. Left unchecked, this heat would destroy the machines within minutes. The processors would throttle, then fail, then melt into useless slag.

So we cool them. We cool them constantly, **at enormous expense**, using systems that have themselves become marvels of engineering. Cold water circulates through pipes that wind among the server racks, absorbing heat and carrying it away. In evaporative cooling systems—still the most common approach—that water eventually becomes steam, rising through cooling towers and dispersing into the air. It is gone. A medium-sized data center can consume 300,000 gallons of water per day. A large one can drink over five million gallons daily—the water usage of a town of 50,000 people.

In 2024, Google's single largest data center, located in Council Bluffs, Iowa, consumed one billion gallons of water. That is enough to supply the residential water needs of the entire state of Iowa for five days. One building. One year.

One billion gallons. This is the machine that answers your questions.
This is the cost you do not see.

• • •

For two decades, data center energy consumption grew relatively slowly. Engineers achieved remarkable efficiency gains that offset the explosive growth in digital traffic. Between 2010 and 2018, global data center workloads tripled while energy consumption increased by only six percent. It was a quiet triumph of optimization—better chips, smarter cooling, more efficient software. The industry could credibly claim that the digital economy was decoupling from its physical footprint. Then came artificial intelligence.

On November 30, 2022, a San Francisco startup called OpenAI released a chatbot named ChatGPT to the public. Within five days, it had one million users. Within two months, it had one hundred million—the fastest adoption of any technology in human history. By 2025, ChatGPT was serving 700 million weekly active users and processing over 2.5 billion queries per day. And ChatGPT is only one product from one company in what has become a global arms race to deploy artificial intelligence. The computational demands of AI dwarf anything the industry had previously encountered. A traditional Google search consumes approximately 0.0003 kilowatt-hours of electricity. A ChatGPT query consumes roughly ten times that amount—and can consume fifty times more for complex queries with long inputs or outputs.
The queries that once took fractions of a second now require sustained computation across thousands of specialized processors. The models that power these systems required months of training on clusters consuming megawatts of power continuously.

The efficiency gains that had masked two decades of growth evaporated almost overnight. According to the Lawrence Berkeley National Laboratory, U.S. data centers consumed approximately 176 terawatt-hours of electricity in 2023—roughly 4.4 percent of total national electricity consumption. By 2028, that figure is projected to reach 325 terawatt-hours at minimum, and could exceed 580 terawatt-hours under aggressive AI adoption scenarios. The upper estimate represents twelve percent of projected U.S. electricity consumption—more than the steel, aluminum, and chemical manufacturing industries combined. The same explosion is happening with water. U.S. data centers directly consumed 17 billion gallons of water in 2023. By 2028, that figure could double—*or quadruple.*

And direct cooling water is only part of the story. The electricity that powers data centers must itself be generated, and thermoelectric power plants—which still provide the majority of U.S. electricity—consume enormous quantities of water for cooling. When you account for this indirect water consumption, the total footprint multiplies by a factor of twelve. We are building an industry that drinks like agriculture and burns like manufacturing, and the costs are no longer invisible.

Data centers are not distributed evenly across the landscape. They cluster in particular places, for particular reasons, and their concentration creates particular problems. Northern Virginia is the capital of the cloud. Seventy percent of the world's internet traffic passes through servers in four counties surrounding Washington, D.C.—Fairfax, Loudoun, Prince William, and Fauquier. The region hosts more than 300 operational data centers, with dozens more under construction. The concentration is a historical accident that became self-reinforcing: the early internet backbone ran through the area, attracting the first data centers; their presence attracted more connectivity; more connectivity attracted more data centers. Today, Loudoun County alone contains more computing capacity than most nations.

The consequences of this concentration are now impossible to ignore. In 2023, all data centers in Northern Virginia collectively consumed nearly two billion gallons of water—a 63 percent increase from 2019. Loudoun Water, the county's water authority, has had to rely increasingly on potable water supplies to feed the data centers, diverting water that would otherwise serve residential and agricultural needs. The electrical grid strains under the load. Dominion Energy, the regional utility, is planning $5.2 billion in new transmission infrastructure, including hundreds of miles of new power lines, simply to keep pace with data center demand.

But Virginia is not unique. Similar concentrations are emerging wherever the economics of computation align: cheap electricity, available water, favorable tax treatment, and proximity to major fiber-optic routes. In Iowa, data centers cluster along the Missouri River, drawing both power from wind farms and water from the aquifers below. In Arizona, they spread across the Sonoran Desert, tapping the Colorado River system even as the American Southwest enters its worst drought in 1,200 years. In Oregon, they line the Columbia River

Gorge, competing for hydroelectric power with aluminum smelters and the endangered salmon runs that depend on managed river flows. Each location makes a certain sense in isolation. Collectively, they represent a pattern of resource arbitrage that is becoming increasingly difficult to sustain. The data centers go where the resources are cheapest. They consume those resources. The resources become less cheap. And the industry begins looking for the next place to extract what it needs. It is the logic of extraction, applied to electricity and water in the service of computation. It is the cloud touching down on Earth, and leaving a mark.

• • •

The communities that host data centers are beginning to calculate what they have traded, and whether the exchange was fair. The industry's promises were seductive: tax revenue, construction jobs, the prestige of hosting critical digital infrastructure. In exchange, localities offered tax incentives, expedited permitting, and access to resources. But the ledger is proving more complex than the initial pitch suggested. Data centers employ very few people relative to their footprint and resource consumption. A facility covering twenty acres and consuming fifty megawatts of power might employ fifty workers once construction is complete. Compare this to a manufacturing plant of similar size, which might employ several hundred. The jobs that remain are specialized—network engineers, security personnel, HVAC technicians—and often filled by workers who commute from outside the community.

The tax revenue, too, has proven less generous than promised. Many states and localities offered aggressive incentives to attract the industry, exempting data centers from sales taxes on equipment, property taxes on facilities, and various other levies. In some jurisdictions, the net fiscal impact of data centers is now being questioned. "They don't really bring any jobs, the taxation levels were lowered to the point where they don't really bring a fiscal benefit to the city, and they require an enormous amount of power," the planning administrator of Chandler, Arizona, told journalists. "We struggle with being able to tell the story that this is a communal benefit." Meanwhile, the costs are becoming harder to ignore. Residential electricity rates are rising in data center corridors as utilities invest in new generation and transmission capacity. Aquifers are declining. Noise from cooling equipment disturbs neighboring homes. Power lines march across landscapes that were once farmland or forest.

In some communities, data centers are consuming more than a quarter of the local water supply.

And so the backlash has begun.

According to a 2025 analysis by Data Center Watch, $64 billion in data center projects across the United States have been blocked or delayed by local opposition since May 2024. In Virginia alone, $900 million in projects have been killed outright, and $45.8 billion face significant delays. Across 28 states, researchers identified 142 activist groups organizing against data center development. The opposition crosses political lines. In Warrenton, Virginia, more than 500 residents—including the actor Robert Duvall—packed a town council meeting to oppose an Amazon data center. In Chandler, Arizona, the city council voted unanimously against a 422,000-square-foot facility despite personal lobbying from former Senator Kyrsten Sinema. In Peculiar, Missouri, a group called "Don't Dump Data in Peculiar" successfully pressured officials to amend zoning laws to prohibit data centers entirely. In Prince William County, Virginia, every town council member who supported a controversial data center project subsequently lost reelection.

A nationwide poll found that only 44 percent of Americans would welcome a data center near their home—making data centers less popular than gas plants, wind farms, or even nuclear facilities. When one Maryland developer proposed converting an abandoned mall into a data center, more than 20,000 people signed a petition against it. The industry that once flew under the radar is now a central issue in local politics across America. And the question it raises is simple but profound: Can the communities that host our digital infrastructure afford its physical costs?

This book is about that question—and about an answer that might seem, at first, impossible. The crisis described in these opening pages is real. The energy demands of artificial intelligence are growing faster than our ability to supply them sustainably. The water consumption of data centers threatens communities and ecosystems. The communities that host digital infrastructure are pushing back, and their grievances are legitimate. The current trajectory is unsustainable by any honest measure. And yet the demand for computation is not going to diminish. Artificial intelligence is not a fad that will pass. The applications being built on these systems—medical diagnosis, climate modeling, drug discovery, autonomous vehicles, fraud detection, accessibility tools for the disabled—are too valuable to abandon.

The question is not whether we will keep computing.

The question is how.

The conventional answer is to build more: more data centers, more power plants, more transmission lines, more cooling towers. This is the answer embodied in proposals like OpenAI's Stargate project—multi-gigawatt AI factories that would each consume more electricity than some small countries. It is an answer that assumes the fundamental physics of computation cannot change, and that our only option is to feed the machine more resources than it currently consumes. But there is another answer. It requires changing the machine itself. For sixty years, computation has been performed by pushing electrons through silicon. This is not a law of nature. It is a technological choice—one that made sense when it was made, and one that has delivered extraordinary benefits. But it is a choice that carries inherent physical limitations, limitations that manifest as heat, as power consumption, as water usage, as all the costs described in this chapter. There is another substrate for computation. One that does not generate heat in the same way. One that does not require water for cooling. One that operates not at the nanosecond timescales of electron-based switching, but at the femtosecond timescales of light.

That substrate is photons. And the technology built upon it is called photonic computing. **This is the subject of this book.**

In the chapters that follow, we will examine the full scope of the impossible problem—the physical, political, and strategic dimensions of our current crisis. We will then explore the physics of light as a medium for computation, and the engineering advances that have made photonic computing practical. We will look at the infrastructure being built to deploy this technology, and the economic and geopolitical implications of a transition from electrons to photons. The story is ultimately hopeful. The crisis is real, but it is not insoluble. The demand for computation can be met—not by consuming ever more resources, but by changing the way we compute.

We can keep computing. We can do it sustainably. *We can do it now. But first, we must understand what we are up against.*

The cost of a question is only the beginning.

Light is faster than electricity.
That fact has consequences.

CHAPTER 2

The End of Moore's Law

In 1965, Gordon Moore made an observation that would shape the next sixty years of human civilization.

Moore was the director of research and development at Fairchild Semiconductor, one of the pioneering companies of what would become Silicon Valley. He had been asked to write a brief article for Electronics magazine predicting where the semiconductor industry might be heading. Looking back at the short history of integrated circuits—the first had been demonstrated only seven years earlier—Moore noticed a pattern. The number of components that engineers could fit onto a single chip was doubling roughly every year. If that trend continued, he speculated, integrated circuits would eventually contain thousands of components, then tens of thousands, then more.

It was a modest prediction, offered with appropriate hedging. Moore was not claiming to have discovered a law of physics. He was describing an empirical trend in manufacturing capability, driven by improvements in lithography, materials science, and engineering ingenuity. He later revised his estimate to a doubling every two years. But the observation took on a life of its own. It became known as Moore's Law, and the semiconductor industry organized itself around it as if it were a commandment from on high.

For six decades, the industry delivered on that commandment with remarkable fidelity. In 1970, transistors measured approximately 12,000 nanometers across. By 2000, they had shrunk to 130 nanometers. By 2020, leading-edge chips were manufactured at 5 nanometers. Each generation packed more transistors onto the same area of silicon, enabling more computation at lower cost. The smartphone in your pocket contains more processing power than the computers that guided Apollo 11 to the moon—and costs a tiny fraction of what those room-sized machines did.

This exponential progress was not an accident. It was not even, strictly speaking, a prediction that happened to come true. It was a self-fulfilling prophecy, a roadmap that the entire industry followed because following it was profitable.

Chipmakers invested billions in research and development because they knew their competitors were doing the same. Equipment manufacturers built ever more sophisticated tools because they knew customers would pay for them. Software developers wrote ever more demanding applications because they knew faster hardware was coming. The expectation of continued progress created the conditions for continued progress.

Until it didn't.

There is a reason Moore's observation was never called Moore's Theory, or Moore's Principle. It was always Moore's Law—a term that carries the implication of something immutable, something beyond human control. But unlike the laws of thermodynamics or gravity, Moore's Law was never written into the fabric of the universe. It was an engineering achievement, and engineering achievements eventually run into physics. The physics that now constrains the semiconductor industry operates at scales almost too small to conceptualize. At these dimensions, the classical behavior of matter gives way to quantum mechanics, and quantum mechanics does not care about roadmaps. To understand why Moore's Law is ending, you must understand what happens inside a transistor at three nanometers.

A transistor is a switch. Current flows, or it does not. The gate that controls this flow is, on a modern chip, approximately ten atoms thick. Ten atoms. The transistor itself is approximately fifteen silicon atoms across. If you could somehow line up the gates of every transistor in a smartphone processor, the line would be shorter than a human hair is wide. At these dimensions, the rules change.

An electron is not a tiny ball that bounces predictably off barriers. An electron is a probability distribution—a smear of likelihood that extends beyond the physical boundaries of the materials that nominally contain it. When a gate is ten atoms thick, there is a meaningful probability that an electron will simply appear on the other side, having passed through the barrier without ever occupying the space within it. This is quantum tunneling. It is not a flaw in the engineering. It is not a problem that better manufacturing can solve. It is the physics of matter at scales approaching the atomic, and it sets a hard limit on how small transistors can become while still functioning as switches. The consequences are not abstract. A transistor that leaks current when it should be off wastes energy—energy that becomes heat. **Heat that must be removed.**

Water that must be evaporated. Power that must be generated. The smaller the transistors, the more of them fit on a chip; the more transistors, the more heat; the more heat, the more infrastructure required to prevent the chip from destroying itself. This is the physics that makes the crisis described in Chapter 1 inevitable. It is not a failure of execution. It is not a failure of investment. It is electrons doing what electrons do when you confine them in spaces measured in handfuls of atoms.

The only way out is to stop using electrons.

And quantum tunneling is not the only problem. As transistors shrink, they pack more tightly together, and the heat they generate concentrates in smaller areas. Modern processors can produce power densities exceeding 100 watts per square centimeter—comparable to the surface of a nuclear reactor. This heat must be removed, constantly, or the chip will destroy itself. But heat removal becomes harder as devices shrink, because the surface area available for cooling decreases faster than the power density increases. The industry has hit what engineers call the "power wall."

There is also the problem of wiring. At advanced nodes, more than 70 percent of a chip's area and a substantial portion of its power consumption are dedicated not to transistors but to the nanoscale wires connecting them. These wires have such high resistance at small scales that they act like tiny heaters, and the delay in signals traveling across them often becomes the primary performance limiter—not the transistor switching speed. The speed of light and the resistivity of metals now dictate performance more than the transistor itself. The industry consensus is converging on a date. Somewhere around 2030, classical silicon chips will reach the end of what physics allows. The road that Gordon Moore first described in 1965 is approaching a cliff.

If the physics of small were the only problem, the industry might find ways to work around it—at least for a while longer. But there is a second crisis unfolding in parallel, and it may prove even more constraining: the economics of extreme miniaturization are breaking down.

Moore's original insight was not just about transistor counts. It was about cost. The subtitle of his 1965 article was "Cramming More Components onto Integrated Circuits," but the point was that each component would become cheaper as more were added. The cost per transistor would fall with each generation, making computation progressively more affordable. This economic dynamic—not just the technical achievement of miniaturization—is what made the digital revolution possible. That dynamic is now reversing. The cost per transistor has stopped falling and, by some measures, has begun to rise. The reason is visible in Veldhoven, a small city in the Netherlands that has become the most critical chokepoint in the global technology supply chain.

Veldhoven is home to ASML, a company that most people have never heard of but that every advanced technology depends upon. ASML manufactures lithography machines—the equipment that prints circuit patterns onto silicon wafers. And ASML has a monopoly on the machines required to produce the most advanced chips in the world.

The monopoly is not the result of anticompetitive behavior. It is the result of physics and decades of relentless investment. ASML's extreme ultraviolet lithography machines—EUV machines—use light with a wavelength of just 13.5 nanometers to etch patterns that are only a few atoms wide. To generate this light, the machines drop microscopic tin droplets into a vacuum chamber, then blast each droplet twice with a high-powered laser 50,000 times per second. The resulting plasma emits the EUV light that makes modern chips possible.

The engineering challenges are almost absurd. EUV light is absorbed by nearly everything, including air, so the entire optical path must be in a vacuum. The mirrors that guide the light must be polished to within a single atom of perfection—if one of ASML's mirrors were scaled to the size of Germany, its tallest bump would be less than a millimeter. The machines contain over 100,000 components sourced from more than 800 suppliers across the globe. They are the size of a bus, require multiple cargo planes to ship, and cost approximately $200 million each. The next-generation high-numerical-aperture EUV machines, now entering production, cost over $350 million. ASML spent thirty years and tens of billions of dollars developing this technology. Intel, Samsung, and TSMC—the three companies capable of manufacturing leading-edge chips—each invested billions more directly in ASML to ensure the technology would exist. No other company has the accumulated knowledge, the supplier relationships, or the financial resources to replicate what ASML has built. China has reportedly invested tens of billions attempting to develop domestic EUV capability. **Experts estimate they are at least a decade behind.**

The result is a concentration of technological power that has no parallel in modern industry. Every advanced processor—every smartphone chip, every AI accelerator, every data center server—passes through ASML's machines. There is no alternative supplier. There is no backup plan.

And the machines are only part of the cost. A leading-edge semiconductor fabrication facility—a "fab"—now costs upward of $20 billion to construct, and the figure is rising with each generation. Only three companies in the world can afford to compete at the cutting edge: TSMC in Taiwan, Samsung in South Korea, and Intel in the United States. TSMC alone manufactures more than 90 percent of the world's most advanced chips.

This concentration has consequences. When the cost of manufacturing leadership becomes prohibitive for all but a few players, the competitive dynamics that drove Moore's Law begin to break down. Investment in the next generation of technology becomes a bet-the-company decision rather than routine business. The pace of progress slows not because the physics is impossible but because the economics no longer support the pace of investment. The consequences of these twin crises—physical and economic—are already visible in the performance of new chips. In the golden era of Moore's Law, each new manufacturing node delivered roughly 40 percent more performance at the same power level, or the same performance at roughly half the power. This was Dennard scaling, the silent partner to Moore's Law that made transistor shrinking worthwhile. Smaller transistors meant faster switching and lower voltage, which meant exponential improvement in the amount of computation that could be performed per watt of electricity.

Dennard scaling ended around 2006. Since then, voltage has remained stubbornly fixed—it cannot be reduced further without causing transistors to become unreliable—and power density has risen with each new generation. The industry has compensated by shifting from faster clock speeds to more processor cores, spreading the work across multiple units operating in parallel. Your computer no longer runs faster than the one you bought five years ago; it runs more things at once. This shift to parallelism has masked the slowdown in fundamental progress, but it has limits of its own. Not all software can be parallelized efficiently. Coordinating multiple cores introduces overhead. And adding more cores does not help when the bottleneck is how quickly a single thread of computation can execute—which is often the case for the AI workloads now driving industry demand.

The performance gains from each new manufacturing node are now measured in single-digit percentages, not the 40 percent improvements of the past. Moving from 5 nanometers to 3 nanometers might yield a 10 percent speed improvement at the same power, or 25 percent power reduction at the same speed. These are valuable gains, but they are not the exponential leaps that built the digital economy.

Meanwhile, the node designations themselves have become largely meaningless. A "3-nanometer" transistor is not actually three nanometers in any meaningful dimension—the gate pitch, the fin pitch, and the metal pitch are all larger. The naming convention has become a marketing exercise, a way for manufacturers to signal continued progress even as the physical reality of that progress diminishes. Industry veterans now refer to the phenomenon with dark humor: the numbers on the roadmap bear less and less relationship to anything happening on the silicon. The industry has responded to these diminishing returns by going vertical—literally. Three-dimensional chip stacking allows manufacturers to pile multiple layers of transistors atop one another, connected by microscopic pillars called through-silicon vias. This approach sidesteps some of the limitations of planar scaling, but it introduces new challenges in heat dissipation, manufacturing yield, and design complexity. It is an engineering workaround, not a solution to the underlying physics.

The industry has also turned to specialized accelerators—chips designed for specific workloads rather than general-purpose computation. Graphics processing units, tensor processing units, and various AI accelerators perform their narrow functions far more efficiently than general-purpose processors. But this, too, is a workaround. It does not solve the fundamental problem; it merely postpones the reckoning.

• • •

In 2022, Jensen Huang, the CEO of Nvidia—the company whose graphics processors have become the workhorses of artificial intelligence—declared that Moore's Law was dead. A few days later, Pat Gelsinger, the CEO of Intel—the company that had been most closely identified with Moore's Law for half a century—countered that Moore's Law was alive and well. The exchange captured the industry's uneasy relationship with its own trajectory. The executives were talking past each other, each using "Moore's Law" to mean something different. But both understood the underlying reality: the exponential progress that had defined their industry for sixty years was coming to an end.

This matters because the demand for computation is not ending. The artificial intelligence systems described in the previous chapter require ever more processing power. The models are getting larger. The training runs are getting longer. The inference workloads are multiplying. And the chips that power these workloads are no longer getting exponentially better.

The industry's current strategy is to paper over the gap with brute force—to build more data centers, deploy more chips, consume more electricity. But brute force has limits, as we have seen. The power and water consumption of data centers are already straining communities and ecosystems. If the chips themselves stop improving at historical rates, the strain will only grow.

What is needed is not another incremental improvement in the existing paradigm. What is needed is a new paradigm altogether—a way of computing that does not rely on pushing electrons through ever-smaller silicon channels. The successor to the transistor will not be a smaller transistor. It will be something else entirely.

That something is the subject of Part II of this book. But before we can turn to solutions, we must complete our survey of the crisis. The problems of energy consumption and physical limits are only part of the story. There are also questions of infrastructure, geography, politics, and national security that must be understood.

The cloud has a weight.
The supply chain has vulnerabilities.

And the communities that bear the burden of our digital infrastructure are no longer willing to **bear it silently.**

The physics was never the problem. The timing was.

CHAPTER 3

The Weight of the Cloud

In January 2025, President Trump stood in the White House alongside the CEOs of OpenAI, SoftBank, and Oracle to announce the largest infrastructure project in American history. It was not a highway system. It was not a power grid. It was not a network of bridges or tunnels or dams or bullet trains.

It was a data center.

The project was called Stargate, and its ambitions matched its name. Five hundred billion dollars in investment over four years. A network of artificial intelligence factories spread across Texas, Ohio, Wisconsin, and eventually abroad. The first site, in Abilene, Texas, would be operational within months, consuming more than one gigawatt of electricity—enough to power 750,000 homes. By the time the project reached full scale, it would require more power than many small countries. The announcement was theatrical, as such things tend to be. The president compared Stargate to the Manhattan Project, invoking the national mobilization that had built the atomic bomb. OpenAI's CEO spoke of hundreds of thousands of jobs and American leadership in artificial intelligence. The numbers were almost too large to process: $100 billion in immediate deployment, nearly seven gigawatts of planned capacity, facilities spreading across multiple states and eventually multiple continents.

What the announcement did not mention was where all that electricity would come from. Or how the power lines would be built to deliver it. Or what would happen to the communities that would host these facilities. Or whether the physical infrastructure of the United States could support such ambitions at all. The cloud, it turns out, is very heavy. And the ground beneath it is beginning to buckle.

• • •

For fifteen years, American electricity demand was essentially flat. This was not supposed to happen. Utilities had built their business models on steady growth—population increases, economic expansion, the natural accumulation of appliances and devices. But beginning around 2008, efficiency gains offset new consumption. LED bulbs replaced incandescents.

Factories optimized their processes. The grid operators learned to predict demand with precision, and the system settled into a comfortable equilibrium.

That equilibrium is now shattering.

In 2023, American utilities submitted their five-year demand forecasts to regulators, as they do every year. The collective projection showed 38 gigawatts of new peak demand expected by 2028. One year later, in 2024, the same utilities submitted updated forecasts. The new projection: 128 gigawatts. In twelve months, the expected growth had more than tripled. The cause was not a mystery. Data centers, particularly those being built for artificial intelligence, were consuming electricity at rates that defied historical patterns.

A single hyperscale AI facility can draw 100 megawatts or more—the equivalent of a small city.

And the facilities were multiplying. According to research firm 451 Group, U.S. data center grid power reached approximately 62 gigawatts in 2025 and is projected to exceed 134 gigawatts by 2030. The Lawrence Berkeley National Laboratory, using more conservative assumptions, projected that data centers could consume between 325 and 580 terawatt-hours annually by 2030—representing between seven and twelve percent of total U.S. electricity consumption.

The higher estimates are staggering. Boston Consulting Group calculated that under aggressive AI adoption scenarios, U.S. data centers could consume more than 1,000 terawatt-hours annually by 2030—roughly a quarter of total American electricity generation in 2023. That scenario may be unlikely, but even the conservative projections represent a fundamental transformation of the American energy landscape. The utilities are scrambling to respond. American Electric Power, which serves eleven states across the Midwest and South, announced that it expects 24 gigawatts of new demand by 2030—of which 18 gigawatts will come from data centers alone. That is five times the company's current system capacity. Dominion Energy, which serves Virginia's data center corridor, has committed $50 billion in infrastructure investment between 2025 and 2029 simply to keep pace with growth.

But building new power plants and transmission lines takes time—far more time than building data centers. A natural gas plant requires five to seven years from conception to operation. A nuclear plant takes a decade or more.

Transmission lines face years of permitting, land acquisition, and environmental review. The data centers are being built in months. The infrastructure to power them is measured in decades. The gap between demand and supply is opening like a chasm, and no one is quite sure how to bridge it.

• • •

Before a new power plant or data center can connect to the electrical grid, it must pass through what is known as an interconnection queue. This is the waiting room of American energy infrastructure—a bureaucratic process by which grid operators evaluate whether the system can handle new loads and what upgrades might be required to accommodate them.

The queue is broken.

According to Lawrence Berkeley National Laboratory, nearly 2,600 gigawatts of generation and storage capacity are now waiting in interconnection queues across the United States—more than twice the total installed capacity of the entire existing power plant fleet. The median wait time from initial application to commercial operation has grown from under two years in 2008 to nearly five years in 2023. In high-demand regions like Northern Virginia, delays can reach seven years or more. Over 70 percent of applications are eventually withdrawn, often because the projects are no longer economically viable by the time approval comes through. The problem is structural. The interconnection process was designed for a different era—one in which new projects trickled in gradually and could be evaluated one at a time. The system used a serial, first-come-first-served approach: each application was studied in sequence, and if any project ahead of you in the queue dropped out, the entire analysis had to be redone. Developers learned to game the system by submitting multiple speculative applications, hoping that at least one would make it through. The result was a queue clogged with phantom projects and a backlog that grows faster than it can be cleared.

In Texas, CenterPoint Energy reported a 700 percent increase in large load interconnection requests between late 2023 and late 2024, growing from one gigawatt to eight gigawatts. Utilities across the country are reporting more gigawatts of data center applications than their historical maximum peak demand. The Federal Energy Regulatory Commission has implemented reforms—shifting to a cluster-based approach, requiring higher deposits and proof of project readiness—but the changes will take years to show results, and the queue continues to grow.

Meanwhile, the physical equipment needed to connect new loads to the grid is itself in short supply. Large power transformers—the devices that step voltage up for transmission and down for distribution—now have lead times of 120 weeks or more. Some orders stretch to 210 weeks, more than four years. The North American Electric Reliability Corporation has warned that transformer shortages pose a risk to grid reliability. Wood Mackenzie estimates that the United States faces a 30 percent supply shortfall for power transformers in 2025.

The shortage is not easily fixed. Transformer production depends on grain-oriented electrical steel, a specialized material with particular magnetic properties. Only one company in the United States—Cleveland-Cliffs—manufactures it domestically, at plants in Pennsylvania and Ohio. Roughly 80 percent of large transformers have historically been imported from Mexico, China, and Thailand. But global demand has surged, and the supply chains that served a stable grid cannot keep pace with explosive growth. The transformers that once cost $400,000 now cost $600,000 or more. Prices have risen four to six times since 2020. And even at those prices, the equipment may not arrive for years. Two data centers in Silicon Valley have been built but cannot begin operations: the transformers to supply them with electricity do not exist.

The costs of this infrastructure crisis are not being borne equally. They are landing on the communities least equipped to absorb them. In the PJM Interconnection, the regional grid operator serving much of the Mid-Atlantic and Midwest, capacity prices for the 2025-2026 planning year increased by $9.3 billion in a single auction. Residential customers in Maryland will see their bills rise by approximately $18 per month. In Ohio, the increase is projected at $16 per month. These are not one-time charges; they are permanent increases in the cost of electricity, driven largely by the infrastructure investments required to serve data centers.

A study by Carnegie Mellon University found that average American electricity bills could rise by 8 percent by 2030 due to data center growth. In Northern Virginia, the epicenter of the data center industry, the increase could reach 25 percent or more. The households paying these higher bills are not the ones benefiting from the AI revolution. They are simply the ones who happen to live near the infrastructure that powers it. Some states are pushing back. Ohio has implemented new tariffs requiring large data centers to guarantee 85 percent minimum usage of their requested capacity—a response to speculative interconnection requests that tied up grid resources for facilities that were never built. The policy caused interconnection requests in one utility's territory to drop from 30 gigawatts to 13 gigawatts overnight, as developers were forced to demonstrate that their projects were real.

Other jurisdictions are going further. In 2024, PJM proposed rules that would require large data centers to bring their own generation—to build power plants alongside their server farms, rather than relying solely on the grid. The proposal reflects a fundamental shift in how regulators view these facilities: not as ordinary customers entitled to service, but as industrial operations that must bear the full cost of their extraordinary demands. The data center operators are responding in kind. Microsoft has contracted to restart a reactor at Three Mile Island, the nuclear plant famous for its 1979 partial meltdown, to secure dedicated power for its facilities. Amazon has signed agreements with nuclear plants in Pennsylvania and Illinois. A company called CloudBurst is building a natural gas pipeline capable of delivering 450,000 MMBtu per day—enough to generate 1.2 gigawatts of electricity—directly to its Texas data center campus. The cloud is becoming its own power system, separate from and competing with the grid that serves everyone else.

This privatization of energy infrastructure has its own consequences. When the largest electricity consumers exit the shared grid, the fixed costs of maintaining that grid are spread across fewer customers. Rates rise for those who remain. The grid becomes a service for those who cannot afford alternatives, while the wealthy build their own systems. It is a familiar pattern in American infrastructure—the decay of shared resources as the privileged opt out—now playing out at the scale of the national power system.

• • •

The constraints facing the data center industry are not merely regulatory or financial. They are physical. The materials required to build digital infrastructure are finite, and the supply chains that deliver them are fragile.

A hyperscale data center is, at its foundation, a building—and buildings require steel and concrete. The American Concrete Association projects that the United States will need approximately one million metric tons of cement for AI data center construction by 2028. That is in addition to the concrete required for housing, roads, bridges, and all the other demands of a growing economy.

The data centers are not replacing other construction; they are adding to it. Beyond the shell, data centers require specialized equipment with long lead times and limited suppliers. Uninterruptible power supply systems, power distribution units, generators, switchgear—each component passes through supply chains that were built for steady demand, not exponential growth. Some items have lead times exceeding six months. Fiber optic cable manufacturing is controlled by approximately eight suppliers globally, only two of which are in the United States.

The concentration is even more extreme for the servers themselves. The AI accelerators that power modern data centers are manufactured almost exclusively by Nvidia, using chips fabricated almost exclusively by TSMC in Taiwan. A single company's factories, located on an island 100 miles from mainland China, produce the silicon that powers the AI ambitions of every major technology company in the world. The supply chain is not merely concentrated; it is singular. And then there is land. The largest data center facilities require 500 to 800 acres of space, in locations with access to fiber optic networks, reliable power, water for cooling, and willing local governments. Such sites are not infinite. The industry is spreading from its traditional strongholds in Virginia and the Pacific Northwest to new territories: Texas, Arizona, Georgia, the Midwest. But each expansion encounters the same constraints—grid capacity, water availability, community resistance—that drove the industry away from its previous homes.

A 2025 study found that power constraints were extending data center construction timelines by 24 to 72 months. Projects that were once limited primarily by land availability and permitting are now constrained by whether utilities can deliver power at all. The bottleneck has shifted from what can be built to what can be powered. And the bottleneck is tightening.

• • •

Return, for a moment, to Stargate.

The project embodies every tension in the data center industry's trajectory. It promises transformative investment—hundreds of billions of dollars, hundreds of thousands of jobs, American leadership in the technologies that will define the twenty-first century. And it requires transformative infrastructure—gigawatts of power generation, thousands of miles of transmission lines, fleets of transformers that do not yet exist, supply chains that cannot yet deliver at scale. Within days of the announcement, critics pointed out that the project's funding was uncertain, its timelines were aggressive, and its power requirements were essentially impossible to meet through conventional means. Elon Musk, himself no stranger to ambitious infrastructure projects, publicly questioned whether Stargate's backers actually had the money they claimed. The project's defenders countered that the skeptics underestimated American ingenuity, that the investment would flow as promised, that the infrastructure would somehow materialize.

Both sides may be right. Stargate may indeed attract hundreds of billions in investment. It may indeed create significant employment and advance American competitiveness in artificial intelligence. And it may also run headlong into the physical limits that this chapter has described—the interconnection queues that stretch for years, the transformers that take four years to deliver, the grid that cannot expand fast enough to serve it. The broader question is not whether Stargate will succeed or fail. It is whether the paradigm that Stargate represents—building ever larger electronic data centers, consuming ever more electricity, demanding ever more from infrastructure that is already strained—can continue at all. The physics described in the previous chapter suggests that the chips themselves are approaching fundamental limits. The infrastructure described in this chapter suggests that the systems supporting those chips are approaching limits of their own.

Something has to give. Either the demand for computation must be curtailed—an outcome that seems unlikely given the economic and strategic imperatives driving AI development—or the way we compute must change.

The problems are real.

The constraints are binding.

The current path is unsustainable.

But the problems are also, in a sense, clarifying. They define the specifications that any solution must meet. A sustainable computing infrastructure must consume dramatically less power. It must generate dramatically less heat. It must operate within the physical constraints of an electrical grid that cannot expand indefinitely and a planet whose resources are finite.

But the physical constraints described in this chapter are only part of the story. The infrastructure that powers computation is strained. The supply chain that builds it is something else entirely: fragile, concentrated, and vulnerable in ways that no amount of domestic investment can quickly resolve.

That vulnerability has a geography. It runs through a single Dutch company's factories in the Netherlands. It passes through fabrication plants on an island 100 miles from mainland China. It depends on materials and expertise concentrated in a handful of locations that could be disrupted by a single geopolitical crisis.

The cloud is heavy.
The stack beneath it is vulnerable.

And the communities that bear its weight are beginning to push back.

CHAPTER 4

The Vulnerable Stack

In Veldhoven, a small city in the southern Netherlands, a company called ASML manufactures machines that weigh 180 tons, cost up to $400 million each, and require three Boeing 747 cargo planes to transport. Each machine contains more than 100,000 components sourced from over 800 suppliers across the globe. The mirrors inside must be polished to atomic smoothness—deviation from perfection must be measured not in millimeters or micrometers but in individual atoms.

These machines create light that exists nowhere else on Earth. A laser fires 50,000 times per second at droplets of molten tin, vaporizing them into plasma hotter than the surface of the sun. That plasma emits extreme ultraviolet light with a wavelength of just 13.5 nanometers—light that must be captured and directed by mirrors operating in a vacuum, because air would absorb it completely. The light passes through a precise pattern and etches circuits onto silicon wafers, drawing lines 6,000 times thinner than a human hair.

ASML is the only company on Earth that makes these machines. There are no alternatives. Every advanced semiconductor fabricated anywhere in the world—the chips powering AI systems, smartphones, data centers, military systems—requires ASML's extreme ultraviolet lithography equipment. The company holds 100 percent of the EUV market and approximately 90 percent of the broader lithography market. Thirty years of research, tens of billions of euros in development, and an ecosystem of 5,000 specialized suppliers have produced what may be the most consequential monopoly in modern industry.

Chris Miller's Chip War, published in 2022, documented how this monopoly emerged and why it matters. Miller traced the decades of investment, the geopolitical maneuvering, and the technical achievements that produced ASML's dominance—and the strategic vulnerabilities that dominance creates. His book became essential reading in boardrooms and government offices precisely because it revealed how thoroughly the global economy depends on a single company, in a single city, building machines that no one else can replicate.

Since Chip War was published, the concentration has only intensified. Export controls have tightened. Capacity constraints have worsened. The question Miller's book raised but could not answer—what happens when the chokepoint fails?—is no longer hypothetical. It is the operating environment of every company that depends on advanced computation. This is not a story about corporate dominance. **It is a story about fragility.**

The semiconductor supply chain—the foundation upon which the entire digital economy rests—is concentrated in ways that would have seemed reckless if anyone had designed it deliberately. The machines come from one company in the Netherlands. The most advanced chips are fabricated overwhelmingly by one company in Taiwan. The rare gases and specialized materials flow through chokepoints that can be disrupted by wars, earthquakes, or political decisions. A single failure at any of these points could bring the global technology industry to its knees.

The production of a single advanced semiconductor involves more than 1,000 processing steps and can cross international borders 70 or more times before reaching a customer. Silicon ingots are grown in Japan, sliced into wafers in South Korea, fabricated in Taiwan, packaged in Malaysia, and assembled into devices in China. The process takes four to six months, requires uninterrupted clean-room operations lasting weeks, and depends on specialized knowledge concentrated among fewer than 50,000 engineers worldwide. At the center of this global network sits Taiwan Semiconductor Manufacturing Company. TSMC fabricates approximately 64 percent of the world's contracted semiconductors and 92 percent of the most advanced logic chips—those produced at process nodes of 7 nanometers and below. The company's customers include Apple, Nvidia, AMD, Qualcomm, and virtually every other major technology firm that designs chips but does not manufacture them. In 2024, Taiwan's total semiconductor output reached $165 billion, a 22 percent increase from the previous year, and the island's share of global advanced chip production continued to grow.

Taiwan is an island approximately the size of Maryland, located 100 miles from mainland China. The Chinese government claims Taiwan as its territory and has not renounced the use of force to achieve what it calls reunification.
The Taiwan Strait, the body of water separating the island from the mainland, carries half of the world's container ships and 80 percent of the largest vessels

by tonnage. Military exercises, blockade scenarios, and the possibility of invasion are not abstract concerns; they are the subject of detailed planning by governments on both sides of the Pacific. The economic consequences of disruption would be staggering. Bloomberg Economics estimates that a Chinese blockade of Taiwan would reduce global economic output by 2.8 percent in the first year alone—approximately $2.7 trillion in lost production. Economists at Johns Hopkins calculate that a blockade scenario would immediately disrupt 90 percent of advanced chip production. It would cost an estimated $350 billion and take three years to replace Taiwan's semiconductor capacity, assuming such replacement were even possible given the concentration of expertise and equipment.

The risks are not limited to military scenarios. In April 2024, a 7.4-magnitude earthquake struck Taiwan, forcing TSMC to evacuate facilities and halt production at plants responsible for manufacturing the world's most advanced chips. The disruption lasted only hours—the company estimated losses of approximately $92 million after insurance—but it served as a reminder of how concentrated and vulnerable the supply chain has become. An earthquake, a typhoon, a pandemic, or a political crisis could interrupt the flow of chips upon which the global economy depends.

• • •

The physical infrastructure of chip manufacturing depends on materials whose supply chains are equally precarious. Consider neon gas, which is used in the lasers that perform photolithography—the process by which circuits are etched onto silicon wafers. Semiconductor-grade neon must be refined to extraordinary purity, and before 2022, Ukraine supplied approximately 90 percent of the highly purified neon imported by the United States. The gas is a byproduct of steel production, specifically from older mills equipped with air separation technology, and the former Soviet Union's industrial legacy made Ukraine the world's dominant supplier. When Russia invaded Ukraine in February 2022, two of the three major Ukrainian neon producers—Ingas in Mariupol and Cryoin in Odesa—were forced to halt operations. Neon prices in China quadrupled. Krypton prices in Japan rose by similar amounts. The semiconductor industry's dependence on Ukrainian neon, largely invisible before the invasion, became front-page news. Companies scrambled to diversify suppliers, invest in recycling technology, and build stockpiles. The immediate crisis was averted, but the vulnerability it revealed persists.

The situation with Chinese materials is more deliberate and more dangerous. China produces 98 percent of the world's primary gallium, a metal essential for compound semiconductors and advanced electronics. It controls approximately 68 percent of germanium production and dominates the processing of rare earth elements critical for everything from electric vehicles to defense systems.

In July 2023, China imposed export controls on gallium and germanium, requiring licenses for shipments and detailed end-use documentation. In December 2024, those controls escalated to an effective embargo on exports to the United States.

The export controls were a direct response to American restrictions on semiconductor technology sales to China—part of an escalating trade and technology conflict between the world's two largest economies. But the underlying vulnerability is structural. China spent decades building dominant positions in critical materials, often through industrial policy and subsidies that made Chinese production cheaper than alternatives. Western countries, comfortable with cheap imports, allowed their domestic industries to atrophy. By the time policymakers recognized the strategic implications, the dependencies were deeply entrenched. By mid-2025, gallium prices in Rotterdam had risen more than 150 percent from pre-control levels.

The U.S. Geological Survey estimated that a complete gallium and germanium export ban could reduce American GDP by $3.4 billion, with nearly half the impact falling on the semiconductor industry. China temporarily lifted some restrictions following diplomatic negotiations, but the underlying message was clear: the materials upon which the technology industry depends can be weaponized at any time.

The United States and its allies have not ignored these vulnerabilities. The CHIPS and Science Act, signed into law in August 2022, authorized approximately $280 billion in new funding to strengthen domestic semiconductor manufacturing and research. The law included $52.7 billion in direct support for chip production, with $39 billion in manufacturing subsidies and 25 percent investment tax credits for fabrication equipment. The European Union committed over $100 billion to similar initiatives. South Korea announced plans to invest more than $470 billion through 2047. Japan launched aggressive programs to rebuild its chipmaking capacity.

The money has begun to flow. TSMC is building six new fabrication plants in Arizona, with the first entering high-volume production in late 2025. Intel received $7.86 billion in CHIPS Act funding for facilities in Arizona, New Mexico, Ohio, and Oregon. Samsung is constructing a major fab in Taylor, Texas. Micron secured $6.2 billion for memory chip production in Idaho and New York. By late 2025, the Semiconductor Industry Association counted more than $630 billion in announced investments across 140 projects in 28 states. But the projects have encountered difficulties that reveal the challenges of rebuilding an industrial base that took decades to offshore. TSMC's Arizona facilities faced delays attributed to labor disputes, cultural clashes between Taiwanese managers and American workers, and the simple reality that building semiconductor fabs is extraordinarily difficult. Samsung's Texas project, originally scheduled to begin production in late 2024, was pushed back repeatedly. Intel's ambitious Ohio facility, once expected to open in 2025, is now scheduled for 2030. The delays are measured not in months but in years.

More fundamentally, the reshoring effort addresses geography without solving the underlying physics. A fab in Arizona still requires ASML's lithography machines—there is no American alternative. It still requires the specialized gases, chemicals, and materials that flow through fragile global supply chains. It still depends on expertise that exists primarily in Taiwan and South Korea. The CHIPS Act may eventually produce a more distributed manufacturing base, but it cannot eliminate the chokepoints that define the industry's structure.

The **Relaxed Lithography** ™ approach described later in this book offers a different path. If photonic devices can be fabricated on mature process nodes—180 nanometers, 90 nanometers, geometries that dozens of foundries worldwide can produce—then the ASML chokepoint becomes irrelevant. The strategic calculus changes. The supply chain constraints that currently limit who can build advanced computing systems would no longer apply. This is not a theoretical advantage. It is a structural one—the kind of advantage that matters when export restrictions tighten, when geopolitical tensions escalate, when the single point of failure that Miller documented finally fails.

Industry analysts estimate that even with aggressive investment, the United States might produce approximately 10 percent of the world's advanced memory chips by 2035, up from less than 2 percent today. That is progress, **but it is not independence.**

The dependency on Taiwan for the most advanced logic chips—the processors that power AI systems—will persist for years. The dependency on ASML will continue indefinitely. The materials controlled by China will remain strategic vulnerabilities unless entirely new supply chains can be built. Taiwan's dominance of semiconductor manufacturing is sometimes described as a "silicon shield"—the idea that the island's irreplaceable role in chip production deters Chinese aggression because disrupting Taiwan would devastate China's own economy and technological ambitions. The theory has a certain logic. China imports approximately 60 percent of its chips from Taiwan, and Chinese companies depend on TSMC for the advanced semiconductors they cannot produce domestically.

But the shield may be weakening. As Western nations work to reduce their dependence on Taiwan, Taiwan's bargaining position erodes. Every fab that TSMC builds in Arizona or Japan is a fab that does not strengthen the island's strategic value. Taiwanese law prohibits the overseas production of the company's most advanced chips—currently the 2-nanometer process—but each generation of technology eventually becomes available abroad. The shield depends on irreplaceability, and irreplaceability is being systematically undermined. Meanwhile, China is investing heavily in domestic semiconductor capabilities. Chinese equipment manufacturers like Naura and AMEC have captured growing shares of certain market segments, though they remain far behind in the most advanced technologies. In December 2025, reports emerged that Chinese researchers had secretly built a prototype EUV lithography machine using knowledge from former ASML engineers, though experts estimated that producing working chips would require several more years of development. The technological gap is real, but it is not permanent.

The geopolitical trajectory is troubling. Trade restrictions escalate. Export controls multiply. Supply chains fragment into competing blocs. A technology that once represented globalization's triumph—the collaborative production of humanity's most complex manufactured object—is becoming a weapon in great power competition. The integration that made semiconductors cheap and abundant also made them vulnerable. The diversification that would make them resilient also makes them expensive and scarce.

The vulnerability of the semiconductor supply chain is not an accident. It is the accumulated consequence of decades of decisions that optimized for efficiency over resilience, cost over security, globalization over strategic autonomy. No single actor designed the system to be fragile; it emerged from countless choices that each made sense individually but collectively created a structure that a single earthquake, a single war, a single policy decision could disrupt catastrophically. The previous chapters described problems of physics and infrastructure—the heat that chips generate, the limits of shrinking transistors, the grid that strains to power them. This chapter has described a different kind of problem: the concentration of irreplaceable capabilities in locations that are geographically, politically, and strategically exposed. The computing infrastructure of the twenty-first century rests on a foundation that is simultaneously essential and precarious.

The industry's response to these vulnerabilities has been to build more of the same: more fabs, more supply chains, more redundancy within the existing paradigm. That response makes sense as far as it goes. But it does not address the fundamental question of whether the paradigm itself—electronic computation at ever-smaller scales, requiring ever-more-exotic materials and ever-more-concentrated expertise—can deliver the computing capacity that the world will demand. **The problems are compounding.**

The physics limits described in Chapter 2 constrain what transistors can do. The infrastructure limits described in Chapter 3 constrain how many can be powered. The supply chain limits described in this chapter constrain where they can be built and by whom. At every level of the stack, the current trajectory encounters obstacles that money and effort alone cannot overcome. And yet, within these constraints, one company has built an empire.

Any honest assessment of the computing landscape must reckon with Jensen Huang and Nvidia. The CEO of Nvidia recognized, earlier than almost anyone, that parallel processing would become the foundation of artificial intelligence.

His leather jacket has become a symbol—part tech-founder uniform, part rock-star affect—and his keynote presentations draw audiences that rival product launches from Apple. Nvidia's developer conferences are not merely commercial events; they are gatherings of the faithful. Engineers get Nvidia tattoos. They speak of CUDA—the proprietary programming framework that locks developers into Nvidia's ecosystem—with the reverence that earlier generations reserved for Bell Labs or Xerox PARC. Nvidia has built the most valuable semiconductor company in history by mastering the current paradigm—by pushing electronic computation to its limits within the constraints this chapter has described. The company's GPUs power the AI revolution. Its market capitalization exceeds two trillion dollars. Its developer ecosystem numbers in the millions. Jensen Huang is not waiting passively for the future; he is actively shaping it, investing in efficiency improvements, interconnect technologies, and responses to the power and cooling constraints that define the current crisis.

This is the incumbent that any new paradigm must displace. Not a complacent monopolist resting on legacy advantages, but a brilliant, aggressive, technically sophisticated competitor with resources measured in hundreds of billions and relationships with every major technology company on Earth. The question is whether any transition can outrun an incumbent this capable—or whether the physics advantages of alternative approaches are so fundamental that even Nvidia's resources cannot close the gap. The answer to that question is the subject of Part II.

But the stack has one more layer to examine. The data centers, the power plants, the fabrication facilities—they must be located somewhere. And the communities that host them are increasingly unwilling to bear the costs. The backlash against digital infrastructure is growing, and it threatens to constrain the industry's expansion in ways that even its leaders have begun to acknowledge.

CHAPTER 5

The Backlash

On the evening of February 14, 2023, more than five hundred residents of Warrenton, Virginia, gathered at Fauquier High School for what would become a Valentine's Day massacre of a different sort. The town council had scheduled a public hearing on Amazon's proposal to build a 220,000-square-foot data center at the eastern gateway to their community. Among those who rose to speak was Robert Duvall, the Oscar-winning actor who had made his home in the surrounding horse country. Nearly 130 speakers followed, each allocated three minutes to make their case. The hearing stretched past two o'clock in the morning. All but three opposed the project. The council voted 4-3 to approve it anyway.

What happened next defied the conventional wisdom that communities rarely sustain opposition to development once approvals are secured.

Yard signs reading "**Stop Amazon, Stop The Towers**" remained in place for months, then years. Residents formed advocacy groups. They filed lawsuits, requested public records, and began organizing for the next election cycle. When town officials refused to release correspondence between staff and Amazon's representatives, citizens' groups sued—and won at the Virginia Court of Appeals, which ruled that the town must disclose more than 3,000 emails and other documents related to the decision. By November 2024, voters had systematically removed every council member who had supported the data center.

"All of the people who voted for Amazon are gone," one of the newly elected council members observed. "This election was a referendum on Amazon."

The newly elected officials issued a joint statement declaring that "no data center has any place in a small town like Warrenton" and pledging to pursue "whatever course of action is available" to reverse the decision. The statement also promised transparency and accountability—explicit repudiations of the closed-door process that had produced the original approval. Slowly and methodically, the statement noted, "all four Warrenton Town Council members who sold out their constituents to Amazon" had been ushered out of office.

Warrenton is not an outlier. It is a harbinger.

The backlash against data centers has emerged with a speed and intensity that has caught the technology industry off guard. According to Data Center Watch, a research project that tracks opposition across twenty-eight states, community resistance blocked or delayed an estimated sixty-four billion dollars in proposed data center construction between May 2024 and March 2025. By the second quarter of 2025, that figure had grown to approximately one hundred billion dollars in a single three-month period—more than the cumulative total of every quarter since 2023. Twenty projects fell through in ninety days. The pace of opposition was accelerating faster than the pace of construction.

The geography of resistance spans the political spectrum and the national map. In Peculiar, Missouri, population six thousand, the Board of Aldermen amended the city's zoning ordinance to exclude data centers entirely after residents organized against a proposed development. In Goodyear, Arizona, a fourteen-billion-dollar project by developer Tract was withdrawn after activists pressured local authorities to block required rezoning. In Chesterton, Indiana, public forums highlighted community concerns until the developer pulled its application. In Prince George's County, Maryland, officials imposed a 180-day pause on new data center proposals after a petition garnered more than twenty thousand signatures to oppose a project that would have converted an abandoned mall into server farms.

The pattern has become sufficiently pronounced that Data Center Watch now identifies 142 advocacy groups mobilizing across twenty-four states.
On Change.org, the petition platform reported that data center-related petitions surged from a single filing in all of 2024 to 113 in 2025, accumulating roughly fifty thousand signatures. Senator Bernie Sanders of Vermont announced he would push for a national moratorium on AI data center construction to "give democracy a chance to catch up" to an expansion that has proceeded largely without regulation. More than 230 environmental and community organizations signed a letter to Congress requesting a pause on new construction until federal standards are established.

For an industry that had grown accustomed to operating in obscurity, the shift has been disorienting. "Before, local opposition was more of an anecdotal possibility," observed Miquel Vila, an analyst at Data Center Watch.

"Now, it's becoming a core feature of development, in the same way issues like land, energy, and water are taken into account." The complaints against data centers cluster around a constellation of grievances, each reinforcing the others. Water consumption ranks among the most visceral, particularly in regions where drought has made every gallon precious. On May 17, 2024, the City Council of Mesa, Arizona, approved an eight-hundred-million-dollar data center development that would require up to 1.25 million gallons of water daily to keep its servers from overheating. Vice Mayor Jenn Duff cast the lone dissenting vote. "This has been the driest twelve months in 126 years," she said, citing data from the National Oceanic and Atmospheric Administration. "We are on red alert, and I think data centers are an irresponsible use of our water."

The typical data center uses between three and five million gallons of water per day, according to researchers at Texas Tech University's Water Resources Center—roughly equivalent to the consumption of a city of thirty thousand to fifty thousand people. The water is needed primarily for evaporative cooling, a process in which liquid absorbs heat from servers and dissipates into the atmosphere. Unlike industrial processes that discharge wastewater for treatment and reuse, much of this consumption simply vanishes. A study by the Houston Advanced Research Center projected that data centers in Texas alone would consume forty-nine billion gallons of water in 2025, potentially rising to 399 billion gallons by 2030. That final figure would be equivalent to drawing down Lake Mead, the largest reservoir in the United States, by more than sixteen feet in a single year.

The geography of data center expansion has proceeded with apparent indifference to hydrology. According to a Bloomberg analysis, roughly two-thirds of data centers built since 2022 have been located in water-stressed regions. Phoenix, where Maricopa County hosts the nation's second-largest concentration of data centers, sits in a desert where the state has already limited home construction to preserve groundwater. The irony is not lost on residents watching their wells run dry while gleaming server farms consume aquifers that took millennia to fill. In Spain, Amazon's three planned data centers in the Aragon region are licensed to use 755,720 cubic meters of water annually—enough to irrigate more than 570 acres of farmland. "These data centers use water that comes from northern Aragon, where I am," farmer Chechu Sánchez told The Guardian. "They consume water. Where do they take it from? They take it from you, of course."

The opacity surrounding water usage has itself become a point of contention.

In Oregon, the city of The Dalles attempted to block a newspaper's request for Google's water consumption data, triggering a thirteen-month legal battle before releasing the records. When the data finally emerged, it showed that 29 percent of the city's water was flowing to a single Google facility. Lawmakers in at least eight states introduced legislation in 2025 requiring data centers to report their water use. Democratic governors in both California and New Jersey vetoed such bills, citing concerns about imposing rigid requirements on an industry of global importance. The industry's resistance to transparency has not improved its standing with communities that feel they are being asked to sacrifice resources without even knowing how much.

In Tucson, Arizona, the city council voted unanimously in August 2025 to oppose Project Blue, the code name for a proposed $3.6 billion Amazon data center campus on the city's southern edge. The decision came after months of sustained community opposition driven by fears about the project's estimated water consumption and concern over how the cost of intense energy needs might get passed on to local ratepayers. "Is the increase in tax revenue and the relatively paltry number of jobs worth the water?" asked Kathryn Sorensen, a professor at Arizona State University and former water director for Mesa. The question has no easy answer, and increasingly, communities are deciding they would rather not find out.

• • •

Carlos Yanes believes he can identify the moments when global internet activity spikes each night. That is when the sounds from the Amazon data center six hundred feet from his Prince William County, Virginia, home grow most pronounced: the revving of machinery, the whir of exhaust fans cycling to cool servers processing the world's queries. On the morning of February 17, 2025, Yanes and his neighbors in the Great Oak community awoke to a new sound—dozens of Amazon emergency diesel generators blaring after a power outage. The noise, he said, was like "a hair dryer right on your ear." Data center noise emanates primarily from two sources. Cooling systems, which operate continuously to prevent equipment from overheating, generate sound levels ranging from 55 to 85 decibels. Backup generators, which can be the size of tractor-trailers, produce noise comparable to city traffic or jet engines when they activate. For context, the American Speech-Language-Hearing Association identifies 70 decibels as the threshold for safe prolonged exposure; above that level, hearing damage becomes a risk.

The constant drone of a data center may not reach that threshold, but it operates around the clock, replacing the silence of night with an industrial hum that never stops. "It's not necessarily that data center cooling systems have incredibly high decibel ratings—though the generators do," explained Michael, a former engineer leading noise study efforts for a Northern Virginia community group. "It's that they emit noise nearly constantly." The distinction matters. Traditional noise ordinances focus on peak levels, not duration. A data center that technically complies with decibel limits can still transform a quiet neighborhood into something that feels like living beside a highway. The steady background roar accumulates into a quality-of-life assault that residents describe in terms of lost sleep, elevated stress, and the inability to enjoy their own backyards.

Many local noise ordinances do not account for cooling systems mounted on rooftops, which are standard designs for large-scale data centers. Community activists in Northern Virginia have taken to conducting their own noise studies with professional equipment, documenting that facilities regularly exceed county limits—particularly on hot days and nights when cooling demands peak. In Fairfax County, Virginia, regulators responded by requiring data centers to maintain at least 200-foot setbacks from residential properties and mandating noise studies before and after construction. Prince William County is considering new limits of 52 decibels during the day and 47 at night—levels that some existing facilities may find difficult to meet.

The health implications remain understudied, but the concerns are real. A 2025 paper in the journal Environmental Science & Policy called for empirical research on long-term health outcomes in data center host communities, noting that "the public health implications of data center expansion remain largely overlooked" despite the rapid growth of the industry. Data centers generate significant noise pollution primarily from diesel generators and heating, ventilation, and air conditioning systems—the same equipment that also produces air pollutants harmful to nearby communities. The combination of constant low-frequency noise and intermittent generator exhaust creates an environment that some residents describe as industrial in character, incongruous with the residential neighborhoods where many facilities have been permitted.

In Memphis, Tennessee, the consequences have been more acute. In 2024, Elon Musk's artificial intelligence company xAI opened a facility called Colossus in South Memphis to power its Grok chatbot.

The site is located in Boxtown, a predominantly Black community founded by formerly enslaved people who built homes from discarded railroad boxcars. The neighborhood's cancer risk from industrial sources was already 4.1 times the national average, according to a ProPublica analysis. Shelby County receives a failing grade from the American Lung Association for ozone pollution and has the highest rates of children hospitalized for asthma in Tennessee. Into this environment, xAI installed thirty-five methane gas turbines capable of generating 420 megawatts of electricity—equivalent to a medium-sized power plant—without obtaining the required air pollution permits. The Southern Environmental Law Center estimates these turbines emit between 1,200 and 2,000 tons of nitrogen oxides annually, making xAI one of the largest sources of smog-forming pollution in the county. Researchers at the University of Tennessee found that peak nitrogen dioxide concentrations in areas immediately surrounding the facility increased by 79 percent after operations began.

Five days after Grok 4 became one of the most powerful language models in the world, Alexis Humphreys, a twenty-eight-year-old Boxtown resident, experienced her first severe asthma attack in fifteen years. "It felt like my chest was caving in," she told Time magazine. Other residents describe chronic coughs, a persistent smell of gas, and a fear of going outside. "I don't go outside no more," one community member said in a video documenting conditions. "I'm scared to breathe." The NAACP and Southern Environmental Law Center filed notice of intent to sue xAI under the Clean Air Act. More than two thousand public comments were submitted opposing the company's belated permit application, with nearly every comment in opposition. "What's happening in Memphis is a human rights violation," said KeShaun Pearson, executive director of Memphis Community Against Pollution. "Elon Musk and xAI are violating our human right to clean air and a clean, healthy environment."

If water and noise are local grievances, electricity costs have become a regional crisis. The infrastructure required to power data centers must be paid for somehow, and increasingly, that cost is showing up on residential utility bills.
In the PJM Interconnection, the regional grid operator serving thirteen states from Illinois to New Jersey, wholesale electricity costs for the 2025-2026 planning year increased by more than 500 percent compared to three years earlier. The amount PJM agreed to pay to secure capacity from power plants—ensuring they are available when electricity use rises—soared from $2.2 billion in late 2022 to $14.7 billion in 2024, then climbed another 9 percent to $16.1 billion in 2025.

The independent watchdog that monitors PJM identified the primary culprit: "The current conditions in the capacity market are almost entirely the result of large load additions from data centers, both actual historical and forecast." Those capacity prices are ultimately passed down to household electricity bills.

In Maryland, residential customers are seeing increases of approximately $18 per month. In Virginia, where Dominion Energy serves the densest concentration of data centers on Earth, prices rose 13 percent in 2025 compared to the previous year. In New Jersey, increases reached 21 percent. In Ohio, the increase is projected at $16 per month. These are not temporary spikes; they are structural increases in the cost of electricity, driven by infrastructure investments that would not be necessary but for the data center explosion.

A Harvard Law School paper found that Virginia and Maryland ratepayers are on the hook for billions of dollars in transmission infrastructure being built to accommodate data center growth—not residential customers. "Without reforms," the researchers wrote, "consumers will be paying billions of dollars for regional infrastructure that is designed to address the needs of just a few of the world's wealthiest corporations. Residential ratepayers are not causing PJM to plan new transmission." The cost of transmission upgrades in the PJM region reached $4.3 billion in 2024 alone, including $1.3 billion for thirty-seven projects in Ohio. These costs are passed on to electricity customers through higher rates.

The political consequences have arrived with remarkable speed. In November 2025, Abigail Spanberger won Virginia's gubernatorial race after promising to make data centers "pay their own way and their fair share" of rising electricity costs. In New Jersey, governor-elect Mikie Sherrill pledged to declare a state of emergency over electric bills on her first day in office. In Georgia, Democrats ousted two Republican incumbents from the state's utility regulatory commission, with one candidate explicitly tying price increases to data center expansion. Georgia Power had raised monthly bills six times in two years, to an average of $175; the company's proposal to spend $15 billion expanding capacity—primarily to serve data centers—had become a political liability.

An Associated Press poll from October 2025 found that electricity bills are a "major" source of stress for 36 percent of American adults. In data center corridors, that stress has found a target.

The industry's defenders point to the tax revenue that data centers generate—and they are not wrong that the facilities pay substantial sums. But the ledger has grown more complicated as communities examine what they traded to attract that revenue. According to Good Jobs First, a nonprofit watchdog that tracks economic development incentives, at least thirty-six states have crafted subsidies specifically for data center projects. Texas will forego over one billion dollars in tax revenue in 2025 through sales tax exemptions, including exemptions on the electricity data centers consume. The state's Comptroller now estimates the program will cost $1.7 billion in 2030 alone and $9 billion between 2025 and 2030. Virginia, home to the largest data center market in the world, abates nearly one billion dollars annually in state and local sales and use taxes—without disclosing which companies benefit or how much they receive.

Over the past decade, data centers have been exempt from paying $2.7 billion in Virginia alone. Illinois lost $370 million in 2024, up from $10 million just four years earlier, an increase of 3,600 percent. The combined cost across states with substantial programs now exceeds several billion dollars annually, and the figures are rising rapidly. "We know of no other form of state spending that is so out of control," the Good Jobs First researchers concluded. The programs typically have no caps on how much any facility or company can benefit, and no limits on how much any state can lose in a given year. Virginia's data center tax exemption constitutes 80 percent of all economic incentive spending in the state. A 2024 study by the Virginia Joint Legislative Audit and Review Commission found that "like most economic development incentives, the data center exemption does not pay for itself" when considering lost revenue.

The jobs promised have often failed to materialize in numbers that match the facilities' footprints. A 2017 U.S. Chamber of Commerce report found that during construction, data centers average 1,688 workers. Once operational, they employ only 157 permanent staff. A 250,000-square-foot data center might employ twenty-five direct workers and another twenty-five contractors. "That building gets finished," observed Terry Rephann, a regional economist at the University of Virginia, "and then what you have is a data center that really doesn't, per square foot, employ a lot of people."

Georgia's legislature passed a bill in 2024 to pause the state's data center incentive program for review. Governor Brian Kemp vetoed it. In Ohio, the legislature voted to eliminate the state sales tax break—an early sign of data centers' emerging political vulnerability—to help pay for a broader income tax cut. Governor Mike DeWine preserved it, writing in a veto message that Ohio needs the exemption to compete "with other states for technology jobs and capital investment." The race to the bottom continues, even as communities question whether they are winning or losing.

The backlash against data centers has scrambled traditional political alliances in ways that have surprised observers on both sides. Republican officials, traditionally friendly to business development, have raised concerns about tax incentives that benefit the world's wealthiest corporations while straining local resources. Democratic officials, typically focused on environmental and labor issues, have found themselves vetoing transparency requirements and defending subsidies for Big Tech. The opposition crosses party lines because the grievances are fundamentally local: rising utility bills affect Republican and Democratic households equally; noise pollution does not respect voting patterns; aquifer depletion is bipartisan.

In Warrenton, the voters who removed pro-Amazon council members included conservatives concerned about property values and progressives worried about environmental impacts. In Peculiar, Missouri, the group "Don't Dump Data in Peculiar" united residents across the political spectrum. In Prince William County, the lawsuits challenging the PW Digital Gateway project—a $24.7 billion data center complex that remains tied up in court—were filed by Citizens for Fauquier County, a group whose membership spans the ideological range of rural Virginia. The industry has begun to respond. A relatively new trade group, the National Artificial Intelligence Association, has distributed talking points to members of Congress and organized local data center field trips to better pitch voters on their value. Meta has taken out advertising campaigns to sell communities on the economic benefits of hosting facilities. Amazon released a study claiming its data centers actually drove down electricity prices by introducing a massive new customer that helps spread the cost of maintaining existing infrastructure. The argument has not convinced residents watching their monthly bills rise.

The moratoriums are spreading. In Georgia, eight counties and cities passed temporary bans on data center development in September 2025 alone. DeKalb County extended its moratorium while commissioners debated regulations that would restrict facilities to industrial areas. St. Charles, Missouri, imposed a one-year ban after residents strongly opposed a project later linked to Google. St. Louis is considering its own pause while updating zoning codes that predate the digital age. In San Marcos, Texas, city council members moved to pause a $1.5 billion proposal after residents voiced concern that the development would create too much demand on energy in an area already vulnerable to brownouts. Atlanta prohibited data centers within a twenty-two-mile radius of the Belt-line overlay district. More than one hundred counties and cities have passed temporary moratorium, zoning limits, or new environmental rules since 2023.

The industry's political response has been mixed. Public Citizen, a consumer advocacy organization, published a report recommending that governments consider temporary moratoriums on new data center development until baseline protections are implemented. The report also called for federal regulation under bulk power market reliability standards, mandatory load flexibility requirements, and a prohibition on federal preemption that would allow states and localities to maintain their own zoning regulations. The Trump administration, by contrast, has moved in the opposite direction—reportedly preparing an executive order that would block states from creating their own AI rules and could be used to override local zoning restrictions on data center construction.

A nationwide poll found that only 44 percent of Americans would welcome a data center near their home—making the facilities less popular than gas plants, wind farms, or even nuclear reactors. The cloud has descended from abstraction to concrete reality, and the reality is unwelcome.

The phenomenon is not uniquely American. In central Mexico, Microsoft opened a large data center in 2024; residents in the area subsequently complained of more frequent power cuts and water outages lasting days or weeks. In the Netherlands, where data centers already consume an estimated 4 percent of national electricity, the government has imposed restrictions on new construction in certain regions. In Ireland, which hosts a disproportionate share of European data centers, regulators have begun refusing grid connections in areas where power demand threatens stability. The global expansion of digital infrastructure is encountering local resistance everywhere it touches down.

I have watched this backlash unfold with particular attention, because I am trying to build something different in the middle of it. In March 2024, Meta announced plans for a hyperscale data center in Richland Parish, Louisiana—

a facility that will eventually consume more than 100 megawatts of power. The company has deep relationships with state government, virtually unlimited capital, and a playbook refined across dozens of similar projects worldwide.

When I mentioned to advisors that True Photonic was exploring Louisiana for our own facilities, the response was blunt: "Meta is a crusher. They will crush you."

Perhaps. But I have also done the math.

A photonic **Clean Compute Center** occupying two to four floors of a commercial building in New Orleans could match the computational output of Meta's hundred-acre hyperscale facility—at a fraction of the power consumption, a fraction of the water usage, a fraction of the grid impact. No cooling towers evaporating millions of gallons. No diesel generators rattling windows at three in the morning. No transmission lines marching across wetlands that are already disappearing into the Gulf.

If we can demonstrate that capability—if we can show that the computational power of a hyperscale facility can fit inside an office building, powered by a fraction of the grid capacity, requiring no special water allocation—the message to communities like Warrenton and Mesa and Boxtown would be unmistakable: **there is another way**.

The era of hyperscale extraction is not inevitable.

The Photonic railroad is opening, and everyone currently riding the silicon rail is invited to switch tracks. That is why the backlash described in this chapter matters to me not just as context but as opportunity. The communities rejecting *data centers* are not opposed to *computation*. They are opposed to a particular mode of computation—one that takes without giving, that externalizes costs onto people who never agreed to bear them, that treats water and electricity and silence as inputs to be consumed rather than resources to be shared.

Photonic computing offers a different bargain. The physics that makes it possible—*light* instead of *electrons*, *femtoseconds* instead of *nanoseconds*, *transparency* instead of *resistance*—also makes it clean. A technology that does not generate the heat does not need the water to dissipate it.

A technology that does not strain the grid does not raise the bills of families who never asked to subsidize it. A technology that does not require rare earths cannot be weaponized by export controls from Beijing. The January 2024 rare earth restrictions that sent gallium prices soaring would have been irrelevant to photonic switches. The materials in the Poovey Stack are globally abundant. The manufacturing processes use mature equipment available from dozens of suppliers. The strategic vulnerabilities described in the previous chapter simply do not apply.

This is not idealism. *It is physics*, **translated into** *business strategy*.

The communities that have said no to hyperscale data centers might say yes to something that solves their problems rather than creating them. The backlash is not an obstacle to photonic computing. It is the market signal that photonic computing was designed to answer.

The crisis is now complete.

In the preceding chapters, we have traced the impossible problem from its origins to its current manifestations. We have seen how the demand for artificial intelligence is driving energy consumption to levels that strain electrical grids and deplete water resources. We have seen how the physics of shrinking transistors is approaching fundamental limits, ending the exponential progress that sustained the digital revolution for sixty years. We have seen how the infrastructure required to power computation—the transformers, the transmission lines, the generating capacity—cannot expand fast enough to meet exploding demand. We have seen how the supply chain that builds the chips themselves is concentrated in locations that are geographically precarious and politically vulnerable. And now we have seen how the communities that host this infrastructure are pushing back. The sixty-four billion dollars in blocked or delayed projects is not merely a business problem for developers. It is a signal that the current trajectory is unsustainable not just physically, not just economically, but politically. The social license to build data centers—the implicit consent of communities to host the physical infrastructure of the digital economy—is being withdrawn.

The activists who have organized against data centers are not Luddites opposed to technology. Many use the very services that data centers enable. They shop online, stream video, query chatbots, store their photographs in the cloud.

Their opposition is not to computing but to a particular mode of computing—one that externalizes costs onto communities that derive little benefit, that consumes resources without transparency, that offers tax revenue in exchange for social disruption and then extracts the tax incentives that would have made the exchange worthwhile. Their question is simple: **Is there a better way?** The industry's response has been to build more of the same, faster. More data centers. More power plants. More transmission lines. More cooling towers. The Stargate project promises five hundred billion dollars in investment; other hyperscalers are committing similar sums. The assumption underlying these commitments is that the fundamental architecture of computation cannot change, and that the only path forward is to feed the machine more resources than it currently consumes.

But what if the assumption is wrong?

What if there is another way to compute—one that does not generate the heat that requires water to dissipate, that does not consume the electricity that strains grids and raises bills, that does not demand the exotic materials concentrated in vulnerable supply chains? What if the sixty years of electronic computation that brought us to this crisis were not the end of the story, but a prologue to something different? The crisis has defined the specifications that any solution must meet. A sustainable computing infrastructure must consume dramatically less power per operation. It must generate dramatically less heat per calculation. It must operate within the physical constraints of an electrical grid that cannot expand indefinitely and communities that will not accept indefinite burdens. It must be buildable with materials and equipment that are not concentrated in a single earthquake zone or controlled by a single geopolitical rival.

Those specifications sound impossible. For sixty years, they would have been impossible. The path of computation ran through smaller transistors, more electrons, more heat, more power, more of everything.
But there is another path. It runs not through electrons but through photons—through light itself. And the physics of that path is different in ways that matter profoundly for everything we have discussed. Part I of this book has described the impossible problem. Part II will introduce the nature of light as a medium for computation.

The transition from crisis to solution begins now.

We are not saving computing. We are deciding whether computation remains a tool of expansion or becomes a constraint we manage.

PART II

THE NATURE OF LIGHT

What if the data center that powers artificial intelligence required no cooling towers, no river, and one-tenth the electricity?

CHAPTER 6

What Light Can Do

In 1905, a twenty-six-year-old patent clerk in Bern, Switzerland, published four papers that would reshape humanity's understanding of the physical world. The most famous of these introduced special relativity and the equation $E = mc^2$. But another paper, published in March of that year, addressed a puzzle that had confounded physicists for decades: the photoelectric effect.

The problem was simple to describe and impossible to explain. When light struck certain metals, it ejected electrons from the surface. This much was expected—light carried energy, and energy could dislodge particles. But the details were wrong. Classical physics predicted that brighter light should eject faster electrons, since more energy was hitting the surface. Instead, experiments showed that the speed of ejected electrons depended only on the color of the light, not its intensity. Red light, no matter how bright, could not eject electrons from certain metals at all. Violet light, even when dim, sent them flying.

Albert Einstein's explanation was radical. Light, he proposed, was not a continuous wave but a stream of discrete packets—particles of energy that would later be called photons. Each photon carried a specific quantity of energy determined by its frequency. A single violet photon packed enough energy to knock an electron loose; a trillion red photons did not, because each individual packet was too weak. The intensity of light determined how many photons arrived, but the frequency determined what each photon could do.

The Nobel Committee awarded Einstein the 1921 prize not for relativity but for this insight into the nature of light. It was, in many ways, the more revolutionary idea. Relativity told us that space and time were stranger than we had imagined. The photoelectric effect told us that light itself—the phenomenon humans had used to see, to signal, to measure—was something we had fundamentally misunderstood.

A century later, that misunderstanding has consequences for everything described in Part I of this book. The data centers consuming gigawatts of power, the cooling towers evaporating billions of gallons of water, the transformers straining under unprecedented loads, the communities blocking construction of facilities they cannot afford to host—all of these crises trace back to a choice made in the middle of the twentieth century. We chose to compute with electrons. We could have chosen photons. The choice made sense when it was made. But the physics of electrons and photons are profoundly different, and those differences matter more with each passing year. To understand why light might solve computing's impossible problem, we must first understand what light is—and what it can do that electrons cannot.

• • •

The physics in this chapter took me three years to learn. I am not a physicist. I am not an engineer. I am a businessman who has spent four decades learning to recognize incomplete pictures—and learning that you cannot complete a picture you do not understand. When Del first explained what the Poovey Switch could become, I nodded in the right places but grasped perhaps a tenth of what he was saying. The terms were unfamiliar. The concepts were abstract. I understood that 150 femtoseconds was fast, but I did not understand why it mattered. I understood that photons generated less heat than electrons, but I could not have explained the physics to anyone else.

Dr. Poovey and Del taught me. Patiently, repeatedly, with diagrams on napkins and whiteboards and, eventually, actual technical papers that I read with a dictionary of physics terms open beside me. Mark Newland filled in the computing implications. Dr. Poovey, walked me through the materials science that made his switching material possible.

I learn fast - though the education will never be complete. But I learned enough to hold my own in conversations with researchers who had spent careers in optics, enough to write the white papers and articles that have introduced the Poovey Stack to investors and partners, enough to explain to you, in the pages that follow, what light can do and why it matters for the future of computation. What I am about to describe is not speculation. It is physics—the same physics Einstein elucidated in 1905, extended and refined by a century of research. The applications are new. The principles are as old as light itself.

Begin with the electron.

An electron is a fundamental particle with a property called charge. When electrons move through a conductor, their collective motion constitutes

electrical current. This current can be switched on and off, directed through gates, manipulated to represent the ones and zeros of binary logic.
The transistor—the device that makes modern computing possible—is essentially a valve that controls the flow of electrons through a semiconductor material.

But electrons do not move through conductors the way water flows through a pipe. They interact constantly with the atomic lattice of the material. An electron drifting through copper encounters nuclei, other electrons, vibrating atoms. Each encounter deflects its path, absorbs some of its momentum, converts some of its kinetic energy into thermal energy. This is electrical resistance, and it is not a design flaw that better engineering can eliminate. It is a fundamental property of charged particles moving through matter.

The consequences are inescapable. Every time an electron passes through a transistor, some of its energy becomes heat. Every time a signal travels along a wire, some of its energy becomes heat. Every computation performed by an electronic processor—every multiplication, every comparison, every bit flipped in memory—releases thermal energy into the surrounding environment. The faster the processor runs, the more operations it performs, the more heat it generates. This is not a limitation of current technology that future innovations will overcome. **It is physics.**

In 1961, the physicist Rolf Landauer of IBM quantified the minimum possible energy cost of computation. Even in a theoretically perfect computer operating at room temperature, he showed, erasing a single bit of information must release at least 2. - 75 × “10^{-21}” joules of heat—a quantity now known as the Landauer limit. This is the thermodynamic floor below which computation cannot go, regardless of how clever the engineering becomes. Modern processors operate roughly a million times above this limit, meaning enormous room exists for improvement in principle. But the practical improvements have slowed, and the gap between theoretical minimum and actual performance has stubbornly refused to close.The reason is not that engineers have run out of ideas. The reason is that electrons, by their nature, convert energy to heat through every material they traverse. The resistance can be reduced but not eliminated. The switching can be optimized but not made lossless. **The heat will always come.**

Now consider the photon.

A photon is a quantum of electromagnetic radiation—a discrete packet of oscillating electric and magnetic fields propagating through space at approximately 299,792,458 meters per second. Unlike the electron, the photon carries no charge. It does not interact with atomic nuclei through electromagnetic forces in the same way. It does not scatter off every atom it encounters. In transparent materials, photons can travel extraordinary distances with minimal loss.

When light passes through a fiber-optic cable, for instance, it can travel hundreds of kilometers before the signal degrades appreciably. The loss in modern optical fiber is approximately 0.2 decibels per kilometer at certain wavelengths—meaning that 95 percent of the light entering the fiber emerges after traveling ten kilometers. The photons are not bouncing off atoms, losing energy with each collision. They are propagating through the material's electromagnetic structure, interacting weakly with the medium, converting almost none of their energy to heat. This property is why the global telecommunications network runs on light rather than electricity. The transatlantic cables that carry internet traffic between continents are not copper wires conducting electrical current; they are glass fibers guiding photons. The choice was made decades ago for simple economic reasons: transmitting information as light requires vastly less energy than transmitting it as electricity, and the infrastructure to amplify and regenerate signals can be spaced much farther apart. The world's data already moves as photons. It is only processed as electrons. The distinction matters because processing—computation—is where the energy crisis originates.

A Google search consumes most of its electricity not in transmitting your query to the data center or sending the results back to your screen, but in the calculations performed on servers once the query arrives. An AI model drawing 100 megawatts is not spending that power moving data through fiber-optic cables; it is spending it running electrons through transistors, generating the heat that must be carried away by cooling systems consuming millions of gallons of water.

What if the processing happened in light? The physics of photonic computation is not speculative. The fundamental operations required for computing—switching signals on and off, combining inputs to produce outputs, storing information for later retrieval—**can all be performed with light.**

The question has never been whether photons can compute, but whether they can compute practically. Consider switching. A transistor in an electronic processor operates by controlling the flow of electrons through a channel. Apply a voltage to the gate, and current flows; remove the voltage, and current stops. The transition between states—from off to on, from zero to one—takes time. In a modern high-performance processor, that transition occurs in approximately one nanosecond: one billionth of a second, a duration in which light travels roughly thirty centimeters.

Optical switching can occur in femtoseconds. A femtosecond is one quadrillionth of a second—one millionth of a nanosecond. In a femtosecond, light travels approximately 0.3 micrometers, about the wavelength of ultraviolet radiation. The switching happens so fast that the term itself requires recalibration of intuition. If a nanosecond were stretched to one second, a femtosecond would be one millionth of that second—one microsecond in the analogy, a duration so brief it would be imperceptible even in the slowed-down timescale.

The speed differential is not incremental. It is not a matter of photons being somewhat faster than electrons. The physics permits operations occurring one thousand to ten thousand times faster than electronic switching allows. This is the difference between a car traveling at highway speed and a spacecraft traveling at orbital velocity. **They are not in the same category.**

Speed alone would be valuable. But the more consequential difference involves energy. An electronic transistor switching states must charge and discharge capacitances—tiny stores of electrical energy built into the device structure. Every transition requires pushing electrons, and pushing electrons requires overcoming resistance and dissipating power. A photonic switch manipulates light, and light passing through transparent materials generates negligible heat. The energy required to perform an optical switching operation can be orders of magnitude lower than the energy required for an equivalent electronic operation. The numbers vary depending on implementation, but the theoretical potential is staggering.

Analyses of photonic computing architectures suggest power reductions of 90 to 95 percent compared to equivalent electronic systems—not through incremental improvements, but through the fundamental physics of using a medium that does not convert energy to heat in the same way.

Ninety percent power reduction means a data center consuming 100 megawatts could perform the same computation while drawing 10 megawatts. It means the cooling infrastructure that dominates operating costs could shrink proportionally. It means the water currently evaporating through cooling towers could remain in aquifers and reservoirs. It means the electrical grid currently straining under data center load could accommodate far more computation without the infrastructure investments now projected to cost hundreds of billions of dollars.

The mathematics are simple.

The physics is established.

The question has always been whether the engineering could make it real.

• • •

The properties that make photons attractive for computation also introduce challenges that electrons do not face. Electrons interact with each other. This interaction is what makes electronic logic possible. A transistor works because electrons flowing through one channel can be influenced by electrons accumulated in a nearby gate. The charged particles affect each other's behavior, and this mutual influence enables computation.

Photons, in their ordinary behavior, do not interact with each other.

Two beams of light can cross paths without affecting each other at all—this is why you can see objects illuminated by multiple light sources without the beams interfering in complicated ways. The property that makes photons energy-efficient—their weak interaction with matter and each other—also makes them difficult to use for logic. How do you build an AND gate, which outputs a signal only when both inputs are present, if the inputs cannot influence each other? This is not an insurmountable problem, but solving it requires care. Certain materials and configurations can make photons interact indirectly—through the medium they traverse, through nonlinear optical effects, through coupling with electronic systems at specific points. The solutions vary in their elegance and their practicality. Some approaches require exotic materials that are difficult to manufacture. Some require intense laser light that reintroduces energy consumption concerns. Some work in laboratory demonstrations but prove fragile or inconsistent in real-world conditions.

The history of photonic computing is substantially a history of attempts to solve this interaction problem—to find ways of making light do logic without sacrificing the advantages that made light attractive in the first place. That history is the subject of the next chapter. The attempts span decades, involve some of the most accomplished physicists and engineers of the twentieth century, and resulted in repeated frustration. The fact that we are returning to the subject now—after so many false starts—reflects not a triumph of hope over experience, but a genuine change in what is possible.

Before turning to that history, though, it is worth understanding more precisely what photons can do when properly harnessed. The physics is not merely different from electronics. It is different in ways that align remarkably well with the specific demands of artificial intelligence—the application that has pushed electronic computing to its breaking point. The computational demands of artificial intelligence are unlike the demands of traditional software. When you run a word processor or a spreadsheet, the computer executes a series of sequential operations: load this value, add it to that value, store the result, check this condition, branch to that instruction. The operations are complex in aggregate but simple individually, and they follow a logical sequence that mirrors how humans think about the problem being solved.

Neural networks—the computational structures underlying modern AI—work differently. A neural network consists of layers of artificial neurons, each connected to neurons in adjacent layers by weighted connections. When an input enters the network, every neuron in the first layer processes it simultaneously, producing outputs that flow to the second layer, where every neuron processes all incoming signals simultaneously, and so on through potentially hundreds of layers. The fundamental operation is not sequential logic but parallel matrix multiplication: taking large arrays of numbers and combining them according to mathematical rules, over and over, billions of times per inference.

Electronic processors can perform matrix multiplication, but they were not designed for it. A traditional CPU processes one or a few operations at a time, cycling through calculations sequentially even when the underlying mathematics is inherently parallel. Graphics processing units—GPUs—were developed for rendering images, a task that involves similar parallel calculations, and were adapted for AI workloads because their architecture aligned better with neural network requirements. Nvidia's dominance in AI hardware stems from this architectural match: their GPUs can perform many calculations simultaneously, accelerating the matrix operations that neural networks require. But even GPUs are electronic devices, subject to all the limitations described above. The electrons still flow through silicon, still encounter resistance, still generate heat. The massive clusters of GPUs that train large language models consume megawatts of power not because matrix multiplication is inherently expensive, but because electronic implementation of matrix multiplication is expensive.

Photons offer a different approach. Light has a property that electrons lack: it can perform certain mathematical operations as a natural consequence of how it propagates through optical systems. When light passes through a lens, the lens performs a Fourier transform on the light—a mathematical operation fundamental to signal processing. When multiple beams of light combine at a detector, their intensities add, performing summation. When light passes through materials with varying optical properties, the pattern of transmission performs multiplication. The operations that dominate AI workloads—the matrix multiplications and accumulations at the heart of neural network inference—can be executed by light passing through appropriately designed optical elements.

This is not a theoretical curiosity. It is a physical reality that has been demonstrated repeatedly in laboratory settings. A beam of light encoded with input values, passing through an optical matrix of controlled transmission elements, emerges with output values that represent the matrix-vector product—the core operation of neural network computation. The calculation happens at the speed of light, limited only by how fast the optical elements can be reconfigured. And because light passing through transparent materials generates minimal heat, the operation consumes a fraction of the energy required by electronic equivalents.

The alignment between photonic physics and AI requirements is not coincidental. Matrix operations are among the few computational tasks for which analog approaches—computing with continuous physical quantities rather than discrete digital bits—can outperform digital electronics. Light, as an electromagnetic wave with continuous amplitude and phase, is naturally suited to analog computation. The resurgence of interest in photonic computing tracks almost exactly with the resurgence of neural networks: both emerged from long periods of relative dormancy within the past decade, and both are flourishing because the computational demands of modern AI reward their particular strengths.

The energy implications deserve concrete illustration.

Training GPT-4, the large language model that powers ChatGPT, reportedly required approximately 50 gigawatt-hours of electricity over several months—enough to power 4,500 American homes for a year. The training run consumed electricity continuously, around the clock, with thousands of GPUs running at near-maximum capacity. The heat generated required constant cooling; the cooling required water and electricity of its own. The total carbon footprint has been estimated at thousands of tons of CO_2, depending on the power sources feeding the data centers where training occurred.

Inference—the process of actually using a trained model to answer questions—consumes less energy per query than training consumed per example. But inference happens billions of times per day across millions of users. The cumulative energy consumption of inference is now believed to exceed training, and the gap widens as AI deployment accelerates. Every query to ChatGPT, every image generated by DALL-E, every video analyzed by automated systems, draws electricity that ultimately becomes heat.

A photonic system performing the same matrix operations at 90 percent lower energy consumption would not merely reduce electricity bills. It would transform the economics and physical footprint of AI deployment.

A facility that currently requires 100 megawatts could deliver equivalent computation with 10 megawatts. The cooling infrastructure sized for 100 megawatts of heat dissipation could serve facilities performing ten times the computation. The water consumption that has made data centers unwelcome in drought-stricken regions could fall to levels that communities might accept. The numbers are not precise—real-world implementations involve compromises, and the exact efficiency gains depend on architectural details that vary between systems. But the order of magnitude is consistent across analyses. Photonic computing does not promise marginal improvements. It promises improvements commensurate with the crisis: reductions in energy and heat sufficient to change whether AI development is sustainable. There is a temptation, when confronting a crisis as severe as the one described in Part I, to grasp at any solution that promises relief. The history of technology is littered with approaches that seemed revolutionary in the laboratory but proved impractical in the field.

Readers who have followed claims of technological transformation before—cold fusion, superconducting computers, quantum computing's imminent arrival—may reasonably suspect that photonic computing belongs in the same category: promising in principle, perpetually five years away in practice.

The skepticism is warranted, and the next chapter will take it seriously. Photonic computing has a history, and that history includes decades of disappointment. Researchers in the 1980s and 1990s believed they were on the verge of optical computers that would supplant electronics. Companies invested hundreds of millions of dollars. Governments funded major research programs. The revolution did not arrive. The electronics industry continued its relentless improvement, and photonics remained confined to communications—moving data between electronic processors but never replacing them.

Understanding why previous attempts failed is essential to understanding why the current moment is different. The failures were not random; they stemmed from specific technical barriers that could not be overcome with the materials, devices, and fabrication techniques available at the time. Those barriers have not all disappeared, but some have, and others have been circumvented through approaches that were not available to earlier researchers.

The physics described in this chapter has not changed. Light still propagates faster than electrons. Photons still generate less heat than charge carriers traversing resistive materials. Optical systems still perform certain mathematical operations with extraordinary efficiency. What has changed is the ability to build practical systems that exploit these properties—systems that can be manufactured at scale, integrated with existing infrastructure, and deployed in the real world where data centers currently consume rivers and burn coal.

Part I of this book described an impossible problem. The energy demands of artificial intelligence are growing faster than our ability to supply them sustainably. The physics of electronic computation imposes fundamental limits on efficiency. The infrastructure required to power current approaches cannot be built fast enough, and the communities asked to host it are refusing. Something must change, or the AI revolution will consume itself.

Part II introduces the nature of the change that is possible. **Light can compute.** *Light can compute fast*—a thousand to ten thousand times faster than electronics.

Light can compute efficiently—consuming a fraction of the energy and generating a fraction of the heat. Light can compute in ways particularly suited to the matrix operations that dominate AI workloads.

The question is not whether light can solve the problem. The question is whether we can build machines that harness what light can do.

That question has been asked before. The answers have been disappointing. But the asking has never stopped, and the conditions for success have never been better aligned.

The history of those attempts—*and what has changed*—is where we turn next.

The question was never
whether light could compute.
The question was whether
anyone could build it.

CHAPTER 7

The History of Photonic Computing

In 1990, the cover of Scientific American featured an image that captured the technological optimism of its era: a beam of light passing through a prism, refracting into a spectrum of colors, above the headline "The All-Optical Computer." The accompanying article, written by researchers from AT&T Bell Laboratories, described a future in which computation would occur entirely in photons, without the heat and limitations of electronic circuits. The authors were not speculating about distant possibilities.

They were reporting on working devices in their laboratory—optical logic gates, optical memory, optical switches—and predicting that practical optical computers were perhaps a decade away. That decade passed. Then another. The all-optical computer did not arrive.

The history of photonic computing is a history of brilliant ideas, substantial investment, and repeated disappointment. Understanding that history is essential to evaluating claims made today, because the physics that makes light attractive for computation has not changed since 1990. What has changed—and what makes the current moment genuinely different—is a combination of factors that previous generations of researchers could not have anticipated. To understand why this time might be different, we must first understand why all the previous times were not.

The dream of optical computing predates the electronic computer itself. In 1819, the French physicist Augustin-Jean Fresnel demonstrated that light behaves as a wave, establishing principles that would eventually enable the manipulation of optical signals with precision. By the late nineteenth century, Lord Rayleigh had shown that optical systems could perform Fourier transforms—mathematical operations fundamental to signal processing—simply by passing light through appropriate lenses. The physicist Ernst Abbe developed diffraction theory that explained how optical elements could manipulate wavefronts. The mathematics of light manipulation was established more than a century before the first electronic computer was built.

The first practical optical computing systems emerged in the 1960s, driven by military applications. The United States Air Force needed to process radar signals faster than electronic systems could manage, and researchers discovered that optical correlators—devices that compared incoming signals against stored patterns using the physics of light interference—could perform the necessary calculations at extraordinary speed. By 1964, the Perkin-Elmer Corporation had built optical processing systems for the National Reconnaissance Office, analyzing satellite imagery using coherent light rather than electronic circuits. The systems were specialized, expensive, and classified, but they demonstrated that light could perform useful computation. The invention of the laser in 1960 accelerated research dramatically. Lasers provided coherent light—waves oscillating in perfect synchronization—which was essential for precise optical manipulation. Through the 1970s, research groups at Bell Labs, IBM, Stanford, and universities across the world explored optical approaches to computing. The work was largely theoretical and experimental, building understanding of what might be possible without yet producing practical devices.

The breakthrough moment appeared to arrive in 1985, when researchers at Bell Labs demonstrated the first optical bistable device small enough to be integrated onto a chip. A bistable device can exist in two stable states—the optical equivalent of a transistor's on and off positions—and can switch between them when triggered by an incoming light signal. The Bell Labs device, called a Self Electro-optic Effect Device or SEED, was fabricated using semiconductor manufacturing techniques similar to those used for electronic chips. It seemed to offer a path from laboratory curiosity to manufacturable technology.
The enthusiasm that followed was intense and, in retrospect, premature. DARPA, the Defense Advanced Research Projects Agency, launched major funding programs for optical computing research. AT&T, IBM, and other technology giants invested heavily. Academic conferences on optical computing drew hundreds of researchers. The 1990 Scientific American article was not an outlier; it reflected a genuine consensus among experts that optical computing's moment had arrived.

The consensus was wrong.

The failures of optical computing in the 1990s were not failures of physics. The physics worked exactly as the researchers understood it. The failures were failures of engineering, economics, and timing—a convergence of practical obstacles that proved far more difficult to overcome than the underlying science had suggested.

The first obstacle was materials. The optical devices that performed logic operations required exotic semiconductor compounds—gallium arsenide, indium phosphide, and more complex alloys—that were expensive to produce and difficult to manufacture at scale. Silicon, the material that had enabled the explosive growth of electronics, was poorly suited to optical applications because it does not naturally emit or efficiently manipulate light. The photonics industry was attempting to build a new manufacturing infrastructure from scratch, at enormous cost, while the electronics industry was refining processes it had perfected over decades.

The second obstacle was integration. An optical computer requires not just optical logic gates but optical memory, optical interconnections, and interfaces to the electronic systems that would provide inputs and receive outputs. Each component presented its own challenges. Optical memory, in particular, proved extraordinarily difficult. The devices that could store information in optical form required constant power to maintain their state, consumed substantial energy, and could not match the density or reliability of electronic memory. A computer without adequate memory is not a computer at all, merely a demonstration.

The third obstacle was the relentless improvement of electronics. Gordon Moore's observation continued to hold throughout the 1990s and into the 2000s. Every year, electronic transistors became smaller, faster, and cheaper. The performance targets that optical computing was aiming for kept receding as electronics achieved them first. An optical device that would have been competitive in 1990 was hopelessly outclassed by 1995; a device competitive in 1995 was obsolete by 2000. The optical computing community was running a race in which the finish line moved faster than they could approach it.

The fourth obstacle was perhaps the most fundamental: optical computing was attempting to replace electronics in applications where electronics already worked well. The computers of the 1990s performed word processing, spreadsheet calculations, database queries, and simple graphics rendering. These tasks were sequential in nature, processing one instruction after another, and electronic processors handled them with increasing efficiency as Moore's Law delivered its annual improvements. Optical computing offered potential advantages in speed, but those advantages were irrelevant if the applications did not require them. A word processor does not need to perform calculations in femtoseconds; it needs to respond quickly enough that the human user does not notice delay. Electronic processors exceeded that threshold decades ago.

By the late 1990s, the major corporate research programs had largely shut down. Bell Labs, restructured and diminished after the breakup of AT&T, no longer had the resources for long-term fundamental research. IBM shifted its optical efforts toward telecommunications rather than computing. DARPA funding dried up as the promised breakthroughs failed to materialize. The researchers who had devoted careers to optical computing dispersed to other fields or retired. The dream did not die—academic research continued at reduced levels—but it faded from the technological mainstream.

The all-optical computer, it seemed, had been a mirage.

• • •

I think about those researchers sometimes—the ones who spent the 1980s and 1990s chasing the optical computing dream, only to watch it evaporate. They were not wrong about the physics. They were not wrong about the potential. They were simply early, in a field where being early and being wrong look identical until history decides otherwise. *Gary Poovey was one of them.*

When Gary developed his switching material, he was working within a research tradition that the broader industry had abandoned. He spent years refining compositions, testing optical properties, pursuing validation—all while the conventional wisdom held that optical computing was a dead end. His team in Israel invested millions of dollars. They proved the physics worked. They achieved switching speeds that the scientific literature said were impossible. And then they hit the same wall that had stopped everyone else.

The physics was validated, but there was no path to a product. No architecture to plug the switch into. No manufacturing strategy. No business model. No one willing to fund an octogenarian physicist pursuing technology that the industry had written off decades earlier.

When Gary walked into our Santa Rosa lab with his Technion report, he was carrying not just his own work but the accumulated weight of a field's unfulfilled promise. The researchers who had given up, the programs that had been shuttered, the careers that had been redirected—all of that history was present in his data.

What Del saw, and what I came to understand, was that Gary had solved the hardest part of a problem that everyone else had abandoned. The switching speed was real. The physics was validated. What remained was everything else: the logic gates, the architecture, the manufacturing strategy, the business model, the narrative that could attract capital and talent to finish what Gary had started.

The history in this chapter is not abstract to me. It is the context in which True Photonic exists. We are not the first to attempt optical computing. We are the latest in a long line, and we are building on foundations laid by researchers who never got to see their work completed.

If we succeed, it will be because we learned from their failures. If we fail, we will join them in the history of promising approaches that could not cross the gap from laboratory to market.

My team intends not to fail.

What emerged from the wreckage of optical computing research was something more modest but commercially viable: optical communications. The physics that made photons attractive for computation—their ability to travel long distances without losing energy, their immunity to electromagnetic interference, their capacity to carry multiple signals simultaneously through wavelength multiplexing—made them even more attractive for moving data between locations.

The telecommunications industry adopted fiber optics enthusiastically, replacing copper cables with glass fibers that could carry thousands of times more information over thousands of times greater distances. By 2000, the backbone of the global internet was optical. The transatlantic cables, the transcontinental links, the connections between major data centers—all ran on light. But the light stopped at the edge of the computer. Signals arrived as photons, were converted to electrons for processing, and were converted back to photons for transmission to the next destination. The conversion happened billions of times per second across the global network, at every router, every server, every switch.

Each conversion consumed energy and introduced delay. The optical network was a series of islands connected by light, with electronic computation happening on each island.

This architecture reflected a practical compromise.

The conversion between optical and electronic signals required devices called transceivers, and transceivers were expensive—but they were less expensive than replacing all the electronic computing equipment with optical alternatives that did not yet exist. The compromise worked well enough that it became entrenched. The telecommunications industry optimized for optical transmission; the computing industry optimized for electronic processing; the interfaces between them were accepted as a necessary cost.

Through the 2000s and into the 2010s, this division of labor held stable. Data centers grew larger, but their internal architecture remained electronic. The servers, the switches, the storage systems—all processed information as electrons. The optical connections between data centers grew faster and more numerous, but the computation that justified those connections remained stubbornly electronic.

Photonics had found its niche, and that niche was **communication** rather than **computation.**

CHAPTER 8

The Breakthrough

Technological history is crowded with claims that did not survive contact with reality. Ideas that were sound in theory but brittle in practice. Demonstrations that impressed in laboratories yet failed under the pressures of scale, cost, and integration. Entire research programs—well funded, well staffed, and well intentioned—that produced compelling results on paper and little else in the world.

Photonic computing occupies a prominent place in that history.

For more than forty years, researchers pursued the promise of computation in light. The motivation was always clear: photons move faster than electrons, generate far less heat, and can carry information with extraordinary efficiency. If computation could be performed optically rather than electronically, many of the limits described in earlier chapters would recede. But again and again, attempts to realize that promise stalled. Bell Labs, IBM, DARPA-funded consortia—none succeeded in producing a practical photonic computer.

The skepticism this history demands is justified. Any claim to have solved the problems that defeated such institutions deserves rigorous examination. Not optimism. Not enthusiasm. **Examination.**

This chapter examines one such claim.

It introduces an approach to photonic computing developed by True Photonic, based on technology invented by physicist Gary Poovey. The discussion is deliberately specific where specificity is possible: naming the architecture, describing the physical principles involved, identifying what has been independently validated and what remains unproven. The tone remains analytical rather than promotional, because the reader who has followed the crisis described in Part I deserves an honest assessment rather than reassurance.

The architecture is known as the Poovey Stack. Its central claim is direct: that it performs logical operations in light, addressing the photon-interaction problem that stalled earlier efforts, without requiring cryogenic temperatures, exotic materials, or leading-edge semiconductor fabrication. If the claim holds, it represents the kind of breakthrough photonic computing has sought for decades. If it does not, it belongs in the long catalog of false starts that preceded it. The stakes justify careful attention.

• • •

The Problem That Would Not Go Away

The central obstacle to photonic computation is deceptively simple. **Photons do not naturally interact with one another.**

This property is precisely what makes light attractive for moving information. Photons passing through transparent materials generate minimal heat, do not suffer resistive losses, and can propagate at extraordinary speed. Two beams of light can cross paths in empty space without affecting each other at all. That is why optical fibers can carry enormous volumes of data without interference.

But computation requires interaction.

An AND gate must produce an output only when two inputs are present. A switch must change state based on a control signal. In electronic systems, this interaction is inherent in the physics. Electrons carry charge. Charge creates electric fields. Electric fields influence the motion of other electrons. A transistor exploits this chain directly: charge accumulated at a gate electrode creates a field that either permits or blocks current flow through a nearby channel. No special conditions are required. **The interaction is intrinsic.**

Photons carry no charge. They create no static electric fields. Two photons passing through the same region of space do not influence each other's trajectories or states in any useful way. This is not a shortcoming of engineering ingenuity; it is a fundamental consequence of the physics governing electromagnetic radiation.

To perform logic optically, something must mediate the interaction.

A material. A structure. A mechanism that allows the presence of one light signal to influence the path or state of another. Over the decades, researchers pursued multiple strategies to address this requirement. Some relied on nonlinear optical materials—substances whose refractive properties change depending on the intensity of light passing through them. In principle, a strong control beam could alter the medium enough to affect a weaker signal beam. In practice, the nonlinear effects were weak, requiring high-intensity light that reintroduced energy consumption problems and demanded materials that were expensive, fragile, and difficult to integrate into scalable systems.

Others adopted hybrid architectures in which light carried data but electronics performed logic. Optical signals were converted to electrical ones, processed by transistors, then converted back to light. These systems improved communication bandwidth but preserved the fundamental inefficiencies of electronic computation. Each optical-to-electrical conversion consumed energy and introduced delay. The approach amounted to using light as a better wire rather than as a computational medium.

Still others achieved optical bistability in exotic compound semiconductors—gallium arsenide, indium phosphide, and related materials whose optical properties could be manipulated by optical inputs. These devices functioned in laboratories and demonstrated genuine switching behavior. But the materials were incompatible with the silicon manufacturing infrastructure that drove cost reduction and scale in electronics. Commercializing such systems would have required building an entirely new fabrication ecosystem at enormous expense.

The efforts failed not because light cannot support computation, but because the specific implementations could not deliver practical systems. The question was never whether photonic logic was physically possible. It was whether it could be made manufacturable, efficient, and scalable under real-world constraints.

• • •

A Different Approach to Interaction.

The Poovey Stack takes a different approach to the interaction problem. Rather than attempting to force photons to interact directly with one another—an approach physics resists—it engineers the environment through which photons travel. The distinction is subtle but decisive. Logical switching is achieved not by inducing photon-photon collisions, but by designing structures whose optical properties can be altered by light in a controlled and persistent way.

An analogy helps clarify the concept.

Consider a railroad switching yard. Trains do not interact directly to determine their paths. One locomotive does not push another onto a different track. Instead, a switching mechanism observes an incoming train and sets the track configuration so that subsequent trains are routed differently. The interaction between trains is mediated by the infrastructure, not by direct contact.

The Poovey Stack operates on an analogous principle, though the physics is entirely different. A control optical signal induces a temporary change in the properties of a structured material layer. That change persists long enough for a subsequent data signal to encounter a modified optical pathway. The data signal is then routed, attenuated, or blocked according to the state established by the control signal. The photons in the two signals never interact directly; their effects are mediated by the engineered structure they traverse.

This approach sidesteps the photon-interaction problem rather than attempting to solve it head-on. By engineering the environment instead of the particles, it opens design space that was inaccessible to earlier efforts focused on direct photon-photon effects.

The term "Poovey Stack" refers to the layered architecture that enables this behavior. Multiple physical phenomena are orchestrated across vertical layers, each contributing a specific function to the switching operation. The word "stack" is used deliberately. As software systems rely on layered abstractions, and networks rely on protocol stacks, the Poovey Stack is a physical stack—an arrangement of materials and structures that collectively enable logic in light. The detailed physics of these layers is examined in the following chapter. Here, the focus is on what the architecture achieves, rather than how it does so.

• • •

Speed, Power, and Manufacturing.

Three specifications matter when evaluating whether a new computing architecture can address the crisis described in Part I: speed, power consumption, and manufacturability.

The first is speed.

Modern electronic transistors switch in approximately one nanosecond—one billionth of a second. This has been sufficient for decades, but the limits are now apparent. Clock speeds have plateaued due to heat dissipation constraints, and data movement between processors and memory has become a dominant bottleneck. The Poovey Stack has been documented at switching times between 150 and 200 femtoseconds. A femtosecond is one quadrillionth of a second. In that interval, light travels roughly 300 nanometers—less than the width of many electronic features on a chip. Measuring events at this scale requires specialized ultrafast metrology at the limits of current instrumentation.

The improvement factor—between five thousand and ten thousand times relative to electronic switching—is not incremental. It represents a transition into a different physical regime.

Speed alone, however, would be meaningless if it came at the cost of proportionally higher energy consumption. The significance of femtosecond switching lies in the physics that enables it. Photons moving through transparent media do not dissipate energy through resistance. They do not collide with an atomic lattice in the way electrons do. The same physics that allows light to traverse hundreds of kilometers of fiber with minimal loss enables computation to occur with dramatically lower energy dissipation.

Which leads to the second specification: **power.**

Electronic processors convert a significant fraction of their input energy into heat. This is not an engineering flaw; it is a consequence of moving charge through resistive materials. Every transistor switch charges and discharges capacitances, and every charge movement dissipates energy. Practical systems operate far above theoretical minimums, and closing that gap has proven extraordinarily difficult.

The Poovey Stack demonstrates power reductions on the order of 90 percent compared to equivalent electronic implementations. The exact figure depends on workload and system configuration, but the magnitude is consistent. Computation that requires tens or hundreds of watts electronically can be performed with a fraction of that power optically.

The implications are direct. Data centers sized for 100 megawatts of electronic computation could deliver comparable output at a fraction of that load. Cooling infrastructure would scale accordingly. Water consumption—driven largely by heat dissipation—would decline in proportion. The strain on grids, transformers, and communities described in earlier chapters would ease.

The third specification is **manufacturing,** and it may be the most consequential.

Unlike leading-edge electronic processors, the Poovey Stack does not require rare earth materials or advanced fabrication nodes. It can be manufactured using mature semiconductor processes—technology nodes that are widely available and no longer dependent on the extreme ultraviolet lithography systems that define the semiconductor industry's most fragile choke-points.

This reflects a fundamental difference between electronic and photonic devices. Electronic speed depends on shrinking dimensions; smaller transistors switch faster because electrons traverse shorter distances. Photonic devices operate on the scale of light's wavelength. Structures much smaller than that scale do not improve performance. Designing at sizes compatible with light's physics allows fabrication at nodes that are widely accessible and well understood.

The strategic implications are significant. A computing architecture that does not depend on a handful of advanced fabrication facilities, concentrated in a small number of regions, represents a different class of infrastructure asset. It can be manufactured domestically, sourced from multiple suppliers, and scaled without competing for the world's most constrained manufacturing capacity.

• • •

Transport Versus Logic.

Readers who follow technology news may be aware of companies working on "photonic computing" or "optical processors." Understanding what the Poovey Stack claims requires distinguishing it from these efforts.

The photonic computing work that has received the most attention and funding in recent years addresses data transport rather than computation itself. Companies like Lightmatter and Ayar Labs have developed impressive technologies for moving data optically between electronic processors. Their systems reduce the energy consumed by data movement, which is a genuine contribution—data movement accounts for a significant and growing fraction of total system power in modern computing.

But these systems do not perform computation in light.

They move data in light and compute in electrons. **The distinction matters.**

The transport companies connect electronic processors with optical pathways. Data leaves one chip as photons, travels through waveguides or fibers, and arrives at another chip where it is converted back to electrons for processing. The computation itself—the logical operations, the arithmetic, the AI inference—still occurs in **transistors.**

The Poovey Stack makes a different claim: that computation itself—the logical operations that constitute processing—**occurs in photons.**

This is the claim that previous photonic computing efforts failed to deliver.

It is the claim that, **when validated**, will address the fundamental energy crisis rather than merely improving data movement.

The distinction is not between good and better versions of the same thing; it is between addressing a symptom and addressing a cause. The transport companies and True Photonic are not competitors in the usual sense. Their technologies address different parts of the system and could potentially be complementary. But for the reader trying to understand whether photonic computing can solve the crisis described in Part I, the distinction matters enormously. Faster data movement does not reduce the power consumption of computation.

Faster computation—computation that **generates less heat** by operating through **different physics**—addresses the problem at its source.

The gap between transport and logic is not unbridgeable. Our NIBLE BUS—documented in U.S. Patent Publication 20240313860, — represents an architecture that crosses this divide: an all-optical system for managing data transfer through photonic switches, steering components, and buffer memories. Unlike transport systems that merely move data between electronic processors, the NIBLE BUS brokers signals using only light signals, performing the data management functions that computing requires without converting to electronic form. A companion publication, U.S. Patent Publication 20240411203, documents a purely optical NAND logic gate—the universal building block from which all Boolean operations can be constructed. These are not theoretical proposals but documented implementations with sufficient specificity to receive USPTO publication, the standard that indicates an invention has been reduced to practice. The publications establish that all-optical data handling and all-optical logic have progressed from laboratory curiosity to formal intellectual property—searchable in public databases, available for examination by anyone who wishes to verify the claims.

• • •

Validation and Limits

Claims require validation.

The switching speeds described above have been independently measured at Technion University in Israel using pump-probe spectroscopy, the standard technique for characterizing ultrafast optical phenomena. The tests used telecommunications-wavelength lasers and detection equipment capable of resolving the infrared signals involved. The observed switching behavior matched theoretical predictions, with measured times in the 150–200 femtosecond range. This validation matters. Allowing interested parties to measure their own performance claims has misled industries before. Independent confirmation establishes that the observed behavior is not an artifact of measurement or interpretation.

What has also been demonstrated is manufacturability at the device level. Working optical switches have been fabricated on silicon wafers using commercial foundry processes. These are physical devices, not simulations. They demonstrate compatibility with the manufacturing infrastructure required for scale.

What has not yet been demonstrated is a complete photonic processor. The transition from individual devices to integrated systems introduces challenges that have defeated promising technologies in the past: signal routing, thermal effects across arrays, interfaces that preserve efficiency, and yields that make production economically viable. The honest assessment is this: the Poovey Stack has crossed the threshold from theory to validated device-level reality. It has not yet crossed the threshold to system-level proof. The physics is sound. The manufacturing approach is plausible. The integration work remains. History offers no guarantees.

• • •

Why This Time May Be Different

Skepticism is warranted, **but dismissal is not.**

Several conditions distinguish the present moment from earlier eras of photonic computing enthusiasm. AI workloads favor the kinds of operations photonic systems perform efficiently. Silicon photonics manufacturing has matured. The pace of electronic improvement has slowed. Efficiency has become a primary metric rather than an afterthought. And social and regulatory pressures now constrain energy-intensive computation.

The Poovey Stack was designed with these conditions in mind. It does not attempt to replace electronic computing wholesale. It targets the specific classes of computation driving today's infrastructure crisis, while interfacing with electronic systems where appropriate. The promise is not an all-optical computer that replaces everything, but a photonic architecture that makes continued computation possible where electronic approaches are becoming untenable.

The obvious question: *if photonic computing is so promising, why aren't NVIDIA and Intel and Google already doing it?*

The answer is **organizational**, not **technical**. Large companies optimize for incremental improvement along established paths. NVIDIA's market capitalization depends on selling more GPUs; their engineering resources flow toward making GPUs faster, not toward replacing them. Intel has a photonics research division, but it competes internally for resources against the core processor business that generates actual revenue. Google's TPU team is measured on AI training performance, not on reinventing the underlying physics.

The incentive structures of large technology companies favor **evolution**, not **revolution**. Photonic computing is revolution—and revolution is risky, expensive, and threatens existing product lines. The rational strategy for a large company is to wait, watch smaller players prove the technology, and then acquire or copy what works.

This creates a window. The window closes when the technology is proven at scale. Until then, a small company with the right physics, the right team, and the right timing can establish a position that larger players cannot easily displace. **The patents filed now become the toll roads of the future.** The facilities built now become the infrastructure that others must license or compete against.

True Photonic, Inc. exists in that window. The question is whether we can build fast enough to establish the position *before the giants wake up.*

And the story had to be told. Not eventually—**Now.**

Before the window closes.

Before the narrative gets written by someone else.

That is what this book is.

• • •

The **breakthrough**, *if it is one*, lies not in discovering new physics, but in making *known physics operationally useful at scale.*

That is the claim. *The evidence warrants serious examination.*

The next chapters provide it.

PART III
THE ARCHITECTURE

Poovey Switches on Silicon Wafer

NOT
was trivial in electronics.
In photonics, it defeated
a generation of researchers.
Until now.

CHAPTER 9

The Poovey Stack

The architecture is the evidence.

The previous chapter introduced the Poovey Stack's specifications—femtosecond switching speeds, ninety percent power reductions, compatibility with mature fabrication processes. It described what the technology accomplishes without explaining how. For the skeptical reader, that gap is significant. Claims are cheap. Specifications announced in press releases have a long history of evaporating when engineers attempt to build actual products. The all-optical computer promised in 1990 foundered not on its specifications but on the physical mechanisms that were supposed to deliver them.

This chapter examines the mechanism. It describes the physical structure of the Poovey Stack—the specific layers, their compositions, their functions—and explains how those structures enable photons to perform logic. The explanation will be technical but not gratuitously so. The goal is not to replicate a physics journal but to provide enough detail that a serious reader can evaluate whether the approach is plausible, where the uncertainties lie, and what distinguishes this architecture from the approaches that failed in previous decades.

The explanation begins at the foundation—literally. Every semiconductor device rests on a substrate, the material that supports the active structures and provides the physical and thermal environment in which they operate. The choice of substrate is not incidental; it shapes what can be built above it. The Poovey Stack's substrate is unconventional, and its unconventionality is essential to the architecture's function.

• • •

A conventional electronic chip rests on silicon—a thin wafer of crystalline silicon dioxide that has been purified to extraordinary levels and processed through dozens of steps to create the transistors, interconnects, and other structures that constitute the circuit. Silicon has dominated semiconductor manufacturing for six decades because it combines useful electrical properties with manufacturability at scale. It forms a stable oxide that serves as an excellent insulator.

It can be doped with impurities to create the p-type and n-type regions that transistors require. Its crystal structure is well understood, and the processes for growing, cutting, and polishing silicon wafers have been refined over generations of industrial experience.

For photonic devices, silicon presents difficulties. Silicon does not emit light efficiently—a fundamental limitation arising from its indirect band gap that cannot be engineered away without altering the material's basic nature. Silicon absorbs light at the wavelengths most useful for photonics, particularly in the visible and near-infrared range, introducing losses that accumulate as signals propagate through the material. The telecommunications industry chose 1,550 nanometer wavelength partly because silicon's absorption is lower there, but even at this wavelength the material remains far from ideal. And silicon's thermal conductivity, while adequate for electronic devices with their distributed heat generation, becomes a constraint when the goal is to build dense arrays of optical switches that must maintain precise alignment and timing.

Previous photonic computing efforts struggled with silicon's limitations in different ways. Some accepted silicon as the substrate and built switching structures from other materials deposited on top—gallium arsenide, indium phosphide, lithium niobate—creating hybrid devices that inherited silicon's thermal and mechanical properties while attempting to overcome its optical limitations. Others abandoned silicon entirely, fabricating devices on substrates of the III-V semiconductors themselves. Both approaches encountered the manufacturing problem described in Chapter 7: the processes for working with these materials were expensive, low-yield, and incompatible with the economies of scale that had made electronic computing affordable.

The Poovey Stack uses a different foundation: a composite substrate of graphene, hexagonal boron nitride, and sapphire. Each material contributes specific properties that the architecture requires, and the combination creates a platform fundamentally suited to optical rather than electronic operation.

Sapphire—crystalline aluminum oxide—forms the structural base. It is mechanically robust, optically transparent across a wide range of wavelengths from ultraviolet through mid-infrared, and thermally stable at temperatures far exceeding any that the devices will encounter in operation. The material's thermal expansion coefficient is well characterized and predictable, allowing engineers to design structures that maintain dimensional stability across operating temperature ranges. Sapphire substrates are already used in certain specialized semiconductor applications, including LEDs and high-power electronics, and can be produced at the scales and quality levels that commercial manufacturing requires. The material provides a rigid, stable foundation that does not expand, contract, or deform in ways that would disrupt the precisely engineered structures built upon it.

Hexagonal boron nitride—sometimes called "white graphene" for its structural similarity to graphene—serves as an intermediate layer and as an encapsulation material for critical structures. Boron nitride is an exceptional electrical insulator, with a band gap exceeding five electron volts, while being thermally conductive—a rare combination that allows it to channel heat away from active regions without providing pathways for electrical interference. When deposited properly, it forms atomically flat surfaces with roughness measured in fractions of a nanometer, creating the smooth interfaces that the single-atom-thick graphene layers require for optimal performance. And it is transparent to the infrared wavelengths at which the Poovey Stack operates, introducing minimal optical loss even in multilayer configurations.

Graphene itself is the most exotic component and the most critical. A single layer of carbon atoms arranged in a hexagonal lattice, graphene possesses properties that seem almost designed for photonic switching. Its electrons behave as massless Dirac fermions, a consequence of the material's unique band structure, responding to electromagnetic fields with extraordinary speed and minimal energy dissipation. Unlike bulk semiconductors where electrons must navigate complex band structures and scatter frequently against lattice defects, electrons in graphene move through a two-dimensional landscape where scattering is minimized and transit times are measured in femtoseconds.

Graphene can absorb and release photons across a remarkably broad spectrum of wavelengths, from terahertz through visible and into ultraviolet, making it useful for the infrared light that telecommunications and data systems commonly employ. A single graphene layer absorbs approximately 2.3 percent of incident light regardless of wavelength—a seemingly modest figure that becomes significant when multiplied across multiple layers and interaction zones. Its optical properties can be tuned by applying electric fields—a characteristic called electro-optic modulation—allowing external signals to control how light interacts with the material without requiring physical reconfiguration. And when properly integrated with surrounding structures, graphene can serve as the active switching element that enables photons to influence the paths of other photons.

The thermal properties of this substrate stack deserve particular attention. The crisis described in Part I of this book is fundamentally a thermal crisis: electronic processors generate heat that must be removed at enormous expense in energy, water, and infrastructure. Any technology claiming to address that crisis must demonstrate not only that it generates less heat but that what heat it does generate can be managed effectively.

The graphene-boron nitride-sapphire substrate achieves thermal conductivity exceeding **2,000 watts per meter-kelvin** in the plane of the graphene layers. For comparison, copper—the standard material for heat spreaders in electronic systems—conducts heat at approximately 400 watts per meter-kelvin. Aluminum, widely used in heat sinks, manages about 200 watts per meter-kelvin. Silicon itself conducts at roughly 150 watts per meter-kelvin. Diamond, often cited as the ultimate thermal conductor and used in specialized high-power applications, reaches roughly 2,000 watts per meter-kelvin. The substrate matches diamond's thermal performance using materials that can be deposited through industrial processes rather than requiring the extreme pressures and temperatures of diamond synthesis.

This thermal conductivity means that whatever heat the devices do generate spreads rapidly across the substrate rather than accumulating in hot spots. The Poovey Stack generates far less heat than electronic equivalents—a consequence of the switching physics described below—and what heat it does generate dissipates efficiently through a substrate designed for thermal management. The combination addresses both sides of the thermal equation: producing less heat and removing it more effectively.

• • •

The substrate would be useless without the ability to manufacture it reliably at scale. This is where many advanced materials technologies have failed: laboratory demonstrations of remarkable properties that cannot be reproduced in production environments, or can be reproduced only at costs that make commercial deployment impossible. Graphene in particular has suffered from a gap between its theoretical potential and its practical availability. First isolated in 2004 by Andre Geim and Konstantin Novoselov at the University of Manchester—work that earned them the 2010 Nobel Prize in Physics—graphene spent the following decade as a material of extraordinary promise and limited commercial impact.

The original method for producing graphene involved mechanical exfoliation—essentially peeling layers from graphite using adhesive tape. This approach yielded high-quality material suitable for research but was obviously unsuitable for manufacturing. Chemical vapor deposition on copper foils emerged as a scalable alternative, but it required transferring the graphene from the copper growth substrate to the target substrate, a process that introduced defects, contamination, and yield losses. The graphene community spent years developing transfer techniques, each iteration reducing defects incrementally but never eliminating them entirely.

Robert Switzer's contribution to True Photonic addresses this challenge directly. Switzer, through Astera Energy, developed a process for growing graphene directly on sapphire substrates using microwave plasma-enhanced chemical vapor deposition—a technique abbreviated MW-PECVD. The process eliminates the transfer step entirely: graphene grows where it will be used, avoiding the defects and contamination that transfer introduces. And it does so quickly—depositing high-quality graphene in under three minutes, a timeframe compatible with high-volume manufacturing.

Previous approaches to graphene synthesis required hours of processing time, exotic precursor gases, or the problematic transfer steps. MW-PECVD eliminates these bottlenecks. The significance of a three-minute growth cycle extends beyond simple throughput calculations. Short process times mean that manufacturing equipment can produce more substrates per day, reducing capital costs per unit. They mean that process variations—the inevitable fluctuations in temperature, gas flow, plasma power, and chamber conditions that occur in any industrial process—have less time to accumulate into defects. A small variation that causes no problem over three minutes might produce significant non-uniformity over three hours. And short cycles mean that the feedback loop between process engineers and production results is fast enough to enable rapid optimization. A process that takes hours to complete can be run only a few times per day; a process that takes minutes can be iterated dozens of times, accelerating the learning curve that moves technology from laboratory to factory.

The graphene produced through this process exhibits the properties that the Poovey Stack requires: uniform coverage across the substrate area, minimal defects in the atomic lattice, and consistent optical and electronic characteristics from batch to batch. Raman spectroscopy—the standard technique for characterizing graphene quality—confirms single-layer coverage with the characteristic G and 2D peaks that indicate proper crystal structure. The manufacturing process has been validated through production of actual substrates used in device fabrication—not merely through theoretical analysis or small-scale demonstrations.

• • •

Above the substrate, the Poovey Stack builds a vertical structure of functional layers—hence the name "stack." The specific number and composition of these layers varies depending on the device being constructed, but the general architecture supports configurations ranging from five layers for simple switching elements to more than one hundred fifty layers for complex integrated circuits. This scalability is essential: a technology that works only for isolated devices would be a laboratory curiosity, while a technology that can scale to large integrated systems has commercial potential.

The layers serve distinct functions within the architecture. Some contain the active switching structures—regions where light can be controlled by modifying the optical properties of the material. These active layers incorporate graphene or related two-dimensional materials in configurations designed to maximize the switching effect while minimizing parasitic losses. Others serve as waveguides, channels that confine and direct light between active regions. The waveguide layers must be engineered to support the optical modes that carry information, with dimensions and refractive indices calculated to ensure that light propagates efficiently without leaking into surrounding structures or coupling inadvertently between adjacent channels.

Still other layers provide isolation, preventing optical or electrical crosstalk between adjacent structures. In a dense array of switching nodes, signals must remain confined to their intended paths; stray light that couples from one channel to another would corrupt data and introduce errors. The isolation layers use materials with carefully chosen refractive indices to create barriers that light cannot penetrate, while remaining compatible with the manufacturing processes used for the rest of the stack. The arrangement is three-dimensional in a way that conventional planar semiconductor designs are not. Electronic chips are essentially two-dimensional structures; the transistors and interconnects lie in a thin layer near the surface of the silicon wafer, with the substrate itself serving merely as mechanical support and heat dissipation. The Poovey Stack exploits the vertical dimension to achieve density and functionality that would be impossible in a single plane. Light can propagate upward, downward, and laterally through the stack, with switching nodes positioned at strategic intersections to control the flow.

The individual layers are thin—measured in nanometers for some critical structures—but the total stack can extend to substantial thickness. This is a departure from the trend in electronic semiconductors, where the active regions have become progressively thinner as feature sizes have shrunk. In photonic devices operating at infrared wavelengths, the relevant dimension is hundreds of nanometers to match the wavelength of the light being manipulated, and useful structures can span micrometers without penalty. The Poovey Stack takes advantage of this design freedom, building upward in ways that electronic designs cannot.

The vertical architecture also enables a manufacturing approach that differs from conventional semiconductor processes. Rather than building the entire circuit in a single wafer and then dicing it into individual chips, the Poovey Stack can be fabricated in layers that are deposited and patterned sequentially. Each layer can be inspected and characterized before the next layer is added, allowing defects to be identified early rather than discovered only after the entire fabrication sequence is complete. This layer-by-layer approach reduces waste and improves yields—critical factors for commercial viability in an industry where profitability depends on the percentage of manufactured devices that actually work.

• • •

The switching mechanism is the heart of the architecture, the element that distinguishes the Poovey Stack from optical interconnect technologies that merely transport light between electronic processors. Recall the challenge outlined in Chapter 8: photons do not naturally interact with each other. Two beams of light can cross in space without either beam affecting the other—a property exploited every time you see multiple light sources illuminating the same scene without interference. A technology that performs computation in light must find a way to make the presence or absence of one light beam affect the behavior of another—without the direct photon-photon interactions that physics does not permit at practical power levels.

The Poovey Stack achieves this through engineered structures called optical switching tracks. Each track consists of a region where the optical properties of the material can be altered by a control signal, changing how a subsequent data signal propagates through the structure. The control signal is itself optical—a pulse of light at a specific wavelength and intensity—but it does not interact directly with the data signal. Instead, it interacts with the switching structure, changing the structure's state in ways that affect subsequent optical signals passing through.

The mechanism exploits phenomena that occur when light interacts with the piezoelectric and two-dimensional materials in the switching structure. When photons of appropriate energy strike the active region, they induce changes in the material's optical properties—its refractive index, its absorption coefficient, its transmission characteristics—for a brief period. During this window of altered properties, a data signal passing through encounters different optical conditions than it would if the control signal had not arrived.

The effect is binary: *light turns off light, creating the logical switching that computation requires.*

The physics is more nuanced than a simple on-off switch. The switching structure uses an optical channel whose dimensions can be altered by light-induced strain in piezoelectric materials. When the control signal arrives, it triggers a nanometer-scale contraction that changes the channel's cutoff frequency. A data signal at a wavelength near this cutoff—1,600 nanometers in the validated configuration—finds itself suddenly unable to propagate through a channel that was transparent moments before. The effect persists only while the control signal maintains the altered state, then relaxes as the piezoelectric material returns to equilibrium.

The duration of this switching window is measured in femtoseconds. The piezoelectric response and the optical channel dynamics operate on timescales far faster than in conventional semiconductors where electronic switching involves charging and discharging capacitances through resistive pathways—processes that take nanoseconds or longer. This is the origin of the Poovey Stack's femtosecond switching speed. The speed is not achieved through clever engineering of conventional materials; it is a consequence of using materials and mechanisms whose fundamental physics operates on these timescales.

The switching energy is correspondingly low. The optical power required to trigger the switching event is measured in milliwatts at the source, reduced by the optical path to approximately 0.1 milliwatts at the switching structure itself. The control pulses that flip switching states consume microjoules or less, compared to the millijoules or joules required for electronic switching operations. The ratio is roughly a million to one, which translates directly to the ninety percent or greater power reduction that the specifications claim.

A single switching node is not a computer. It is not even a logic gate, in the sense that a transistor is not a logic gate but rather a component from which logic gates can be constructed. The translation from switching elements to computational capability requires additional architecture: ways to combine multiple switches to perform logical operations, ways to route signals between operations, ways to synchronize the timing of inputs and outputs so that data flows correctly through the system.

This translation is the subject of the next chapter, which examines how the Poovey Stack implements the Boolean logic operations that computation requires. Here, the focus remains on the physical mechanism—on establishing that the switching phenomenon is real, that it occurs at the claimed speeds and energies, and that it can be manufactured using processes compatible with commercial production.

• • •

The mechanism has been validated through independent testing at Technion University in Israel, as described in the previous chapter. What that validation demonstrated—switching speeds of 150 to 200 femtoseconds, three to four times faster than any previously published result, and five thousand times faster than electronic switches in current optical networks—confirms that the physics operates as the theory predicts. The cross-correlation measurements established not merely that switching occurs, but that it occurs through the intended mechanism: the signal appeared at the correct position, matched theoretical predictions in shape, and showed switching contrast consistent with practical use in logical operations. The testing also identified areas for future optimization—signal noise that could be reduced through narrower wavelength filtering, and the potential for even faster switching components that equipment constraints prevented exploring fully. The materials and structures have continued to evolve since the Technion validation, with improvements to the switching contrast, signal quality, and manufacturing consistency.

This is the normal progression of technology development: initial validation establishes that the physics works, subsequent engineering refines the implementation. What the validation could not establish—what no measurement technique can establish until the devices are built—is that large arrays of switching nodes will work together as an integrated system. The gap between individual device performance and system-level functionality is where many promising technologies have failed. True Photonic is currently engaged in the engineering work that addresses this gap, moving from validated switches to validated circuits and ultimately to validated processors. The timeline is estimated at eighteen months.

• • •

The characterization process itself deserves mention, because it illustrates both the capability and the current limitations of the technology. Measuring events that occur in 150 femtoseconds requires instrumentation operating at the frontier of what metrology can achieve. The standard oscilloscopes and logic analyzers used to characterize electronic circuits cannot resolve events shorter than a few hundred picoseconds—roughly a thousand times too slow. Even the fastest photodetectors available commercially have response times measured in tens of picoseconds, far too slow to capture the dynamics of femtosecond switching.

The cross-correlation technique used at Technion circumvents these limitations through an elegant physical principle. Rather than trying to detect the switching event directly with a fast detector, the method uses the speed of light itself as the timing reference. Two pulses derived from the same laser source travel different path lengths before reaching the switching structure. By varying one path length with a precision translation stage, researchers can control the relative arrival time of the pulses with femtosecond resolution. A change in path length of 0.03 millimeters corresponds to a time difference of 200 femtoseconds—the stage movement during which the switching signal was observed.

This measurement approach confirms that individual switching nodes operate as designed. What it cannot confirm—what no measurement technique can confirm until the devices are built—is that large arrays of switching nodes will work together as an integrated system. The gap between individual device performance and system-level functionality is where many promising technologies have failed.

• • •

The honest assessment is this: the Poovey Stack represents a physically plausible approach to photonic logic with validated device-level performance and a manufacturing pathway that does not require exotic equipment or processes. The switching mechanism operates on sound physical principles using materials whose properties are well characterized. The substrate technology addresses both thermal management and manufacturability. The architecture scales from individual devices to complex systems through a layer structure that can be fabricated with existing industrial capabilities. And the switching speed has been validated at a world-class research institution using standard ultrafast characterization methods.

What remains unvalidated is integration. The challenge of building working circuits from working switches, working systems from working circuits, and working products from working systems has defeated photonic computing efforts before. The physical principles are not in question; the engineering is. The manufacturing yields that make commercial production viable, the interconnection schemes that route signals without prohibitive losses, the interfaces to electronic systems that must provide inputs and receive outputs—these are the challenges that separate laboratory demonstration from commercial deployment.

The approach to integration that True Photonic has adopted reflects lessons from the semiconductor industry's history. Rather than attempting to build a complete photonic processor in a single leap—an approach that failed spectacularly for several prominent computing startups in recent years—the development proceeds through stages: first simple circuits of a few switches, then more complex circuits, then subsystems, then systems. Each stage validates not only the technical performance but the manufacturing processes, the test methods, and the design tools that production will require. The complete timeline is measured in years, not months, and the uncertainties are real.

But the physics is real as well. The switching speeds are not projections or targets; they are measurements taken at Technion University and documented in validation reports. The power consumption is not an estimate; it is a calculation from measured switching energies. The substrate thermal conductivity is not a theoretical prediction; it is a characterization of fabricated materials.
The foundation of the Poovey Stack is not aspirational; it is demonstrated.
The question is whether what has been demonstrated can be extended, integrated, and scaled to create useful systems. That question can only be

answered through engineering—through the painstaking work of building, testing, debugging, and refining that transforms validated devices into validated products. The work is underway. The outcome is not guaranteed.

• • •

For the reader who has followed photonic computing's history of disappointment, the details in this chapter may provoke skepticism rather than enthusiasm. The materials are different from previous efforts, but materials have been different before. The speeds are faster, but speed claims are easy to make and difficult to verify independently. The manufacturing approach is more practical, but manufacturing approaches always seem practical until they encounter the realities of production at scale.

This skepticism is warranted. The appropriate response to the Poovey Stack is neither acceptance nor rejection but **observation**. The technology makes testable claims. The device-level performance has been verified at Technion using equipment and methods available to any properly equipped ultrafast optics laboratory. The integration stages that follow will succeed or fail in ways that will become evident. The market will not accept photonic computing on the basis of promises; it will accept it on the basis of working products that deliver the claimed advantages in real deployments.

What the architecture demonstrates is that photonic logic is physically possible in a form compatible with commercial manufacturing. This is not a small achievement. The previous chapter reviewed decades of efforts that could not make this demonstration—that required exotic materials available only from specialized suppliers, cryogenic temperatures maintained by expensive refrigeration systems, or fabrication processes that existed only in a handful of research facilities worldwide. The Poovey Stack uses materials that can be produced at scale, operates at room temperature, and has been fabricated through standard foundry processes.

These characteristics do not guarantee success, but they remove barriers that blocked previous efforts.

The next chapter examines how computation emerges from the architecture—how switching nodes combine to form logic gates, how logic gates combine to perform useful operations, and how those operations map to the AI workloads that are driving the infrastructure crisis. The physics of individual switches is necessary but not sufficient; what matters for the crisis described in Part I is whether those switches can be organized into systems that deliver meaningful computation at meaningful scale.

The architecture provides the foundation. The logic provides the capability. And the capability is what the world desperately needs—computation that does not require rivers of water, forests of cooling towers, and the output of power plants dedicated to answering questions. Whether the Poovey Stack delivers on that need remains to be proven. That it offers a credible path—validated at femtosecond timescales in the laboratories of one of the world's leading technical universities—**is what this chapter establishes.**

Forty years of building things taught me one lesson: the best time to start something impossible is before anyone agrees it's possible.

CHAPTER 10

Logic-in-Light

A light-switch is not a thought.

The validated Poovey Switch—operating at femtosecond speeds, consuming microjoules of energy, fabricated through standard foundry processes—represents a physical capability, not a computational one. The previous chapters established that light can turn off light at speeds and energies that dwarf electronic alternatives.

This chapter addresses the more consequential question: **Can that capability be organized into something that actually computes?**

The question is not rhetorical. The history of computing includes numerous switching technologies that worked beautifully in isolation but could not be assembled into useful systems. Magnetic bubble memory switched reliably; it never became a computer. Josephson junctions switched faster than any transistor of their era; they powered no commercial products. The gap between a working switch and a working computer is vast, and many technologies have fallen into it.

What follows is an examination of how the Poovey Stack bridges that gap—how individual switching nodes combine to form logic gates, how those gates combine to perform useful operations, and how those operations map to the AI workloads that are driving the infrastructure crisis described in Part I.
The examination will be technical, because the details matter. It will also be honest about what has been demonstrated and what remains theoretical, because readers deserve to understand where validated physics ends and engineering projection begins.

• • •

Every digital computer ever built—from the room-sized ENIAC of 1945 to the smartphone in your pocket—performs its operations through Boolean logic. Named for George Boole, the nineteenth-century English mathematician who formalized the algebra of logical operations, Boolean logic reduces all computation to combinations of simple binary decisions.

A value is either true or false, one or zero, on or off. Complex calculations emerge from the systematic combination of these elementary choices.

The fundamental operations of Boolean logic are surprisingly few.

NOT inverts a value: if the input is true, the output is false, and vice versa. AND produces true only when both inputs are true. OR produces true when either input is true. From these three operations, all others can be constructed. XOR—exclusive or—produces true when exactly one input is true. NAND—not and—inverts the result of AND. NOR—not or—inverts the result of OR. XNOR—exclusive nor—produces true when both inputs are the same.

Seven operations in total, and from them every calculation a computer performs can be derived.

This is not a simplification for pedagogical purposes; it is a mathematical fact with profound practical implications. A technology that can implement the complete Boolean logic family can, in principle, perform any computation. A technology that cannot implement even one of the fundamental operations is fundamentally limited, useful perhaps for specialized applications but incapable of general-purpose computation. The completeness of the logic family is the threshold that separates a switching technology from a computing technology.

The Poovey Stack implements the complete Boolean logic family. This claim requires substantiation, because previous photonic computing efforts often stumbled on exactly this point. Optical AND gates were demonstrated in the 1990s; optical NOT gates proved far more difficult. Without NOT, there is no NAND; without NAND, there is no functional completeness; without functional completeness, there is no general-purpose computer.

The mechanisms by which the Poovey Stack achieves each logical operation can be described at the architectural level without disclosing the specific material compositions and geometries that constitute True Photonic's intellectual property.

• • •

The NOT gate is the foundation.

In electronic circuits, NOT is trivially implemented: a transistor inverts its input signal as a natural consequence of its switching behavior.

In photonics, inversion is far from trivial. Light does not naturally negate other light. Two photons passing through the same space do not cancel each other; *they simply coexist*. Creating optical NOT required engineering a structure where the presence of an input signal causes the absence of an output signal—where light, in effect, turns off light.

The Poovey Stack achieves optical NOT through the switching mechanism described in the previous chapter. A continuous probe beam passes through an optical channel—a waveguide engineered to transmit light at a specific wavelength. When no control signal is present, the probe beam propagates through the channel and emerges as the output: input absent, output present. When a control signal arrives at the switching structure, it induces the piezoelectric contraction that shifts the channel's cutoff frequency, blocking the probe beam: input present, output absent. The truth table is precisely that of a NOT gate, achieved entirely through optical means.

The probe beam is not the input being negated; it is a carrier signal that the input modulates. This distinction matters for understanding how the architecture scales. The energy required to switch the channel is independent of the energy in the probe beam, allowing weak control signals to gate strong data signals—an optical analog to the amplification that transistors provide in electronic circuits. The architecture avoids the fan-out limitations that plagued earlier photonic logic proposals, where each stage necessarily attenuated the signal until it became unusable.

The AND gate combines two input signals through sequential switching structures. Each input controls a separate switching node, and the nodes are arranged in series along the optical path. The output signal must traverse both switches to reach the output port. If either input is absent—if either switch remains open—the signal is blocked. Only when both inputs are present, triggering both switches to their transmissive state, does the output emerge. The physical arrangement directly implements the logical requirement: output true if and only if both inputs are true.

The OR gate uses a parallel rather than serial arrangement. Two optical paths, each controlled by one input, converge at a common output. If either input is present, its corresponding path transmits, and the output receives a signal. Only when both inputs are absent—both paths blocked—does the output remain dark. The spatial arrangement of the waveguides implements the logical function without requiring any additional control circuitry.

The compound gates—NAND, NOR, XOR, XNOR—emerge from combinations of these primitives. NAND cascades an AND structure with a NOT structure, inverting the AND output. NOR similarly inverts OR. The cascade adds delay—two switching events rather than one—but femtosecond operations cascade into hundreds of femtoseconds, still thousands of times faster than electronic equivalents.

XOR requires particular attention because it is computationally significant: XOR is the fundamental operation of addition in binary arithmetic, the gate that determines whether two bits sum to zero or one before considering the carry.

The implementation in the Poovey Stack uses parallel paths with cross-coupled switching. Each input controls two switching nodes: one that enables its own output path, and one that disables the parallel path controlled by the other input. When input A alone is present, it enables path A and disables path B; the output emerges through path A. When input B alone is present, the mirror situation obtains. When both are present, each disables the other's path; no output emerges. When neither is present, neither path is enabled; again, no output. The truth table of XOR emerges from the geometry without requiring sequential evaluation.

XNOR—the complement of XOR—uses similar geometry with the addition of a NOT stage at the output. The gate produces true when both inputs match: both present or both absent. In cryptographic applications and error-detection circuits, XNOR serves as a bitwise equality comparator, a function that electronic implementations achieve through multiple transistors. The photonic implementation requires no more switching nodes than XOR itself, just different routing of the output signal.

The spatial nature of these implementations deserves emphasis. In electronic circuits, logic gates are switching devices whose inputs and outputs are voltage levels on wires. The physical size of the gate has no direct relationship to its l ogical function; a smaller transistor performs the same logical operation as a larger one, just faster and with less power. In the Poovey Stack, the logic is embedded in the geometry—in the arrangement of waveguides, the positioning of switching nodes, the routing of optical paths.

The architecture is inherently parallel because light propagates through all paths simultaneously, reaching resolution in the time required for a single switching event regardless of the complexity of the logical function being computed.

• • •

Functional completeness is a precise mathematical concept with a surprisingly simple definition: a set of logic gates is functionally complete if every possible Boolean function can be expressed using only gates from that set. The NAND gate alone is functionally complete—any Boolean function whatsoever can be implemented using only NAND gates. The same is true of NOR. The set containing AND, OR, and NOT is also functionally complete. The Poovey Stack, implementing all seven standard gates, exceeds the minimum requirement by a comfortable margin.

Why does this matter? Because functional completeness is the boundary between a technology that can perform specific operations and a technology that can perform any operation. A system limited to AND and OR gates—lacking NOT—can compute only monotonic functions: functions whose output can only increase as inputs increase. Such a system cannot implement subtraction, cannot perform comparison, cannot execute conditional branching. It can do arithmetic, but only addition. It can store data, but cannot search it efficiently. The limitations compound until the system, however fast its individual operations, proves inadequate for practical computation.

The history of photonic computing includes multiple demonstrations of partial logic families—optical AND gates in 1987, optical OR gates using different mechanisms in 1992, optical XOR for specific telecommunications applications in 2001. What these demonstrations shared was an inability to achieve inversion efficiently. The NOT gate, trivial in electronics, required either electronic conversion at some stage—sacrificing the speed advantage of optical processing—or optical mechanisms that introduced unacceptable losses, power consumption, or complexity. The research community produced hundreds of papers describing optical gates that worked in isolation but could not be combined into complete systems.

The Poovey Stack's approach to NOT—using the channel cutoff mechanism to block a probe beam—solves the inversion problem without electronic conversion and without the power penalties of earlier optical approaches. The switching energy is the same for NOT as for AND or OR: approximately 0.1 milliwatts at the switching structure. The speed is the same: 150 to 200 femtoseconds. And because NOT is achieved, all compound gates follow automatically. NAND becomes AND followed by NOT, operating at twice the single-gate delay but still completing in under 400 femtoseconds. The architecture achieves what previous approaches could not: a complete optical logic family operating at consistent speed and power across all operations.

This achievement is now documented in the public record. U.S. Patent Publication 20240411203 describes a purely optical NAND logic gate—the universal building block from which all Boolean operations can be constructed. The publication establishes that the core logical element required for general-purpose computation has been reduced to practice in photonic form, with sufficient specificity to receive USPTO publication. This is not a theoretical proposal but a documented implementation, searchable in patent databases and available for examination by anyone who wishes to verify the claims. The optical NAND represents the culmination of what decades of photonic computing research sought but failed to achieve: a manufacturable, all-optical gate capable of implementing any Boolean function.

I remember asking Gary a question early in our work together. "If we can make light turn off light," I said, *"why can't we just build a computer out of that?"*

Gary looked at me with the patience of someone who had spent twenty-five years thinking about these problems. Then he walked me through what **functional completeness** actually requires—why AND alone is not enough, why OR alone is not enough, why the NOT gate that electronics achieves trivially had defeated optical computing for decades.

"The problem," he said, "was never making light do something.
The problem was making light do everything."

That conversation lasted a couple hours. By the end, I understood why the history described in Chapter 7 was a history of partial solutions—why researchers could demonstrate individual gates but could not build complete systems. The Poovey Stack was designed from the beginning to achieve functional completeness, not as an afterthought but as the primary engineering constraint. Every material choice, every layer configuration, every switching geometry was evaluated against the question: Does this get us to a complete logic family? The optical NAND documented in that patent publication is the proof that the answer is **yes.**

• • •

The reader familiar with computer architecture may be wondering how the Poovey Stack relates to the von Neumann model that has dominated computing since 1945. Named for John von Neumann, who formalized it in a famous 1945 report, the von Neumann architecture defines the structure of virtually every general-purpose computer: a processing unit that performs arithmetic and logical operations, a memory unit that stores both data and instructions, and a control unit that fetches instructions from memory and directs the processor to execute them. The architecture's defining characteristic is the stored-program concept—the insight that instructions themselves can be represented as data, stored in the same memory as the data they manipulate, and modified by the programs they constitute.

The von Neumann architecture is both a triumph and a bottleneck. Its flexibility enabled the software revolution: the same hardware can run word processors, video games, weather simulations, and artificial intelligence simply by loading different programs. But its structure creates a fundamental limitation. The processor and memory are separate, connected by a communication channel of finite bandwidth. Every instruction must be fetched from memory before it can be executed; every data element must be retrieved before it can be processed. The bandwidth of this channel—the rate at which information can flow between processor and memory—bounds the speed of the entire system.

This limitation is called the von Neumann bottleneck, and it has grown more severe as processors have become faster. A modern CPU can execute billions of operations per second, but it cannot fetch billions of instructions per second from main memory. The disparity is managed through elaborate hierarchies of cache memory—small, fast memories that store frequently accessed data close to the processor—but the fundamental mismatch remains. The processor spends substantial time waiting for data, its computational capacity idled by the bandwidth limitation that von Neumann's architecture imposes.

The question for photonic computing is whether it represents a departure from the von Neumann model or an optical implementation of it. The answer is both, depending on how the technology is deployed.

At the system level, the Poovey Stack can implement a von Neumann architecture in optical form. The logic gates described above can be combined to create arithmetic units, control units, and memory interfaces, just as electronic gates are combined in conventional processors. An optical von Neumann machine would retain the stored-program flexibility while operating at femtosecond switching speeds and dramatically lower power consumption.

This configuration represents a direct replacement for electronic computing: same architecture, different physics, better efficiency.

But the Poovey Stack also enables architectures that depart from von Neumann's model in ways that exploit the unique properties of optical processing. Light propagates simultaneously through multiple paths; it does not queue for a shared bus. A photonic processor can be designed with massive parallelism built into its geometry, performing thousands of operations in the time required for a single instruction fetch in a conventional architecture. For workloads that can exploit such parallelism—and the AI workloads driving the current infrastructure crisis are precisely such workloads—the non-von Neumann configuration may prove more valuable than the traditional one.

The architecture also addresses the von Neumann bottleneck through a capability called **Storage-in-Light** ™, or **SIL**. Rather than shuttling data between separate processor and memory components, SIL enables optical storage within the processing fabric itself. Data persists in optical resonant structures that can be read and written at femtosecond timescales, eliminating the round-trip latency of conventional memory access. The processor does not wait for memory because memory is not elsewhere; it is distributed throughout the computational structure, accessible at the speed of light.

The honest assessment is that SIL remains earlier in its development arc than the switching technology. The Poovey Switch has been validated at Technion with measured performance matching or exceeding theoretical predictions. SIL has been designed and analyzed but not yet subjected to equivalent independent validation. The physics is sound—optical bistability and resonant cavity effects are well-characterized phenomena—but the gap between theoretical design and validated performance must be acknowledged. True Photonic's development roadmap positions SIL as a near-term milestone rather than a demonstrated capability.

The infrastructure crisis described in Part I is not caused by computing in general; it is caused by a specific type of computing that has exploded in demand since 2022. Understanding how photonic logic addresses that crisis requires understanding what AI systems actually do when they process your queries—what operations consume the electricity, generate the heat, and drink the water.

The dominant operation in modern AI is matrix multiplication. A neural network processes information by multiplying input vectors against weight matrices, applying activation functions, and passing the results to subsequent layers. A single query to a large language model like GPT-4 involves trillions of multiply-accumulate operations—multiplying pairs of numbers and adding the results—distributed across hundreds of matrix operations involving matrices with billions of elements.

The energy consumption that Part I documented, the water usage that has communities blocking data center construction, the grid strain that has utilities scrambling to build new power plants—all of it traces back to this single mathematical operation performed at unprecedented scale.

Consider what happens when you ask a large language model to summarize a document. The model encodes your query as a vector of numbers—typically 4,096 or more dimensions in current models. That vector is multiplied against a weight matrix with billions of parameters to produce a transformed representation. The process repeats through dozens of transformer layers, each involving multiple matrix multiplications: the attention mechanism alone requires computing query, key, and value matrices, then multiplying queries against keys to compute attention scores, then multiplying scores against values to produce the output. A single forward pass through GPT-4 involves approximately 1.8 trillion multiply-accumulate operations. A conversation with dozens of exchanges multiplies that figure accordingly.

Matrix multiplication has a property that makes it exceptionally well-suited to parallel processing: every element of the output matrix can be computed independently. The entry in row i, column j of the product matrix depends only on row i of the first matrix and column j of the second matrix, not on any other entries in either matrix or any other entries in the output. A processor with sufficient parallel resources can compute all output entries simultaneously, completing the entire multiplication in the time required for a single element.

Electronic processors exploit this parallelism through architectures like GPUs, which pack thousands of simple processing units onto a single chip. But even GPUs face limits: the units must communicate through shared memory, they must be powered and cooled, and the parallelism is bounded by the physical number of units that can be fabricated.

Photonic matrix multiplication exploits a different kind of parallelism. Light passing through an optical structure interacts with all elements of that structure simultaneously—not sequentially, not through time-shared resources, but in a single pass at the speed of light. A matrix encoded in the optical properties of a device—the transmissivities of a grid of switching nodes, for example—multiplies an input vector encoded as light intensities in a single optical transit. The operation that requires thousands of clock cycles in a GPU completes in femtoseconds in a photonic processor.

The Poovey Stack implements matrix multiplication through arrays of switching nodes arranged in a crossbar configuration. The rows encode one matrix; the columns encode the other. Input signals propagate along the rows, interacting with the switching nodes whose states have been programmed to represent matrix weights. The products accumulate along the columns, emerging as output intensities that represent the matrix product. The geometry directly implements the mathematics, with the parallelism inherent in optical propagation replacing the sequential accumulation that electronic processors require.

The specific architectures for photonic AI inference—tensor operations, attention mechanisms, transformer layers, and the other building blocks of modern neural networks—are documented in True Photonic's patent portfolio. The filings establish intellectual property protection across the computational structures that AI workloads require, from basic matrix operations through the complex layer architectures that characterize state-of-the-art models.

The energy advantage is dramatic. A matrix multiplication that consumes millijoules in an electronic GPU consumes microjoules in a photonic implementation—three orders of magnitude difference. The thermal advantage follows directly: less energy dissipated means less heat generated, which means less cooling required. A data center performing the same AI computations photonically rather than electronically would consume a small fraction of the power and virtually none of the water.

The crisis described in Part I is not inevitable; it is a consequence of solving the wrong problem—***optimizing electronic multiplication instead of replacing it.***

The scaling behavior differs fundamentally between electronic and photonic approaches. Electronic matrix multiplication scales linearly with matrix size—doubling the matrix requires roughly double the energy and time. Photonic matrix multiplication, exploiting the inherent parallelism of optical propagation, can scale sublinearly: larger matrices require more optical elements but not proportionally more time, because all elements resolve simultaneously. For the massive matrices that characterize modern AI—weight matrices with billions of parameters, attention matrices spanning thousands of tokens—this scaling difference compounds into substantial advantages.

The mapping extends beyond matrix multiplication to the other operations that AI workloads require. Activation functions—the nonlinear transformations that give neural networks their expressive power—can be implemented through optical nonlinearities in the switching structures. The ReLU activation function, which outputs zero for negative inputs and passes positive inputs unchanged, maps naturally to optical thresholding behaviors. The softmax function, which converts raw scores into probability distributions, involves exponentials and divisions that require more complex optical arrangements but remain achievable within the architecture.

Attention mechanisms, the innovation that enabled transformer models like GPT, involve matrix operations that the same crossbar architecture can perform. The convolutional operations used in image processing map naturally to the spatial arrangement of optical elements—convolution is, mathematically, a sliding dot product, and dot products are what optical crossbars compute. Pooling operations, which reduce spatial dimensions by selecting maximum or average values from regions, can be implemented through optical combining elements. The workloads that are driving the infrastructure crisis are precisely the workloads that photonic processing handles most efficiently.

• • •

The skeptical reader should be asking what has not been demonstrated. The answer is significant: **integrated systems**. The individual switching nodes have been validated at Technion. The logic gate configurations follow from the switching behavior through well-understood optical principles.

The matrix multiplication geometry exploits mathematical properties that are beyond dispute. But a chip that performs useful computation—a photonic processor that accepts real AI workloads and returns correct results—does not yet exist.

The gap is not unique to photonic computing. It is the standard gap that every new computing technology must cross: from validated components to integrated circuits, from integrated circuits to functional processors, from functional processors to commercial products. Intel's first microprocessor, the 4004, required years of engineering between the initial transistor technology and the working chip. The same progression lies ahead for the Poovey Stack, and the uncertainties are real. Integration introduces challenges that do not appear at the component level. Thousands of switching nodes operating simultaneously will generate thermal gradients that could affect timing and signal integrity. Although the Poovey Stack generates far less heat than electronic alternatives, it generates some heat, and that heat must be managed across structures measured in micrometers. The graphene-boron nitride substrate's exceptional thermal conductivity—exceeding 2,000 watts per meter-kelvin—provides the foundation for thermal management, but translating that material property into system-level thermal stability requires careful engineering.

Optical signals propagating through complex geometries will experience losses and crosstalk that must be managed through careful design. A signal traversing multiple waveguides, passing through multiple switching nodes, interacting with multiple control structures, will lose intensity at each stage. The cumulative loss must remain within bounds that preserve signal integrity—the output must be strong enough to register correctly and distinct enough from noise to avoid errors. Previous photonic computing efforts encountered loss budgets that proved unworkable at scale; the Poovey Stack's design addresses loss through shorter optical paths and higher switching contrast, but integration will test whether those theoretical advantages survive practical implementation.

The interfaces between photonic processing fabric and electronic input/output systems must be engineered to preserve the efficiency advantages. Data enters the system as electronic signals from memory or networks; it must be converted to optical form for processing. Results emerge from optical computation; they must be converted back to electronic form for storage or transmission.

These interfaces—optical-to-electrical and electrical-to-optical converters—consume power and introduce latency. If the conversion overhead exceeds the processing savings, the system fails economically regardless of how efficiently the photonic core operates.

Manufacturing yields—the percentage of fabricated devices that actually work—must reach levels that make production economically viable. Semiconductor manufacturing achieves yields exceeding ninety percent for mature processes; new technologies typically start much lower and improve through iterative refinement. A yield of fifty percent means half the wafers produced are scrap— acceptable for research but ruinous for commerce. The Poovey Stack's compatibility with mature fabrication processes provides a starting point higher than exotic technologies would achieve, but the specific yields that commercial production will reach remain to be determined.

These challenges are not insurmountable, but neither are they trivial. The history of semiconductor development includes many technologies that failed during integration despite excellent component performance. True Photonic's approach—progressing through stages from simple circuits to complex systems, validating at each stage before advancing—follows the methodology that successful semiconductor companies have used for decades. But following the right methodology does not guarantee the right outcome.

What can be said with confidence is that the physics supports the claims. Light can perform Boolean logic at femtosecond speeds using the mechanisms described in this chapter. Matrix multiplication can be implemented optically with the energy efficiency that the specifications claim. The operations that AI requires are precisely the operations that optical processing performs most naturally. The question is not whether photonic logic is possible—it demonstrably is—but whether it can be engineered into practical systems at commercial scale.

• • •

The implications of photonic logic, if it can be scaled, extend beyond efficiency improvements to structural changes in how computing systems are built. Electronic data centers are designed around the thermal problem: dense racks of servers, elaborate cooling systems, power distribution networks capable of handling megawatts per building. The architecture reflects the physics of electronic computation—the unavoidable heat generation that accompanies electron flow through resistive pathways.

Photonic computing changes the physics and therefore the architecture. A processor that generates a fraction of the heat does not require the same cooling infrastructure. A facility that consumes a fraction of the power does not require the same electrical capacity. The windowless warehouses and cooling towers that characterize today's data centers may prove to be artifacts of a particular technological era—necessary for electronic computation, unnecessary for optical.

Chapter 12 will explore what this could mean for the physical infrastructure of computing—how photonic data centers might be configured, where they might be located, how they might integrate into urban environments that currently exclude computation as too resource-intensive. But that discussion presupposes that photonic processors can be manufactured, and manufacturing is itself a technical challenge with specific requirements and constraints. The next chapter examines the manufacturing pathway: the specific processes by which Poovey Stack devices are fabricated, the compatibility with existing semiconductor infrastructure that enables commercial production, and the scaling path from current capabilities to volume manufacturing. The logic described in this chapter is meaningless without the ability to build it; the building described in the next chapter validates whether this particular logic can enter the world as physical products.

What this chapter establishes is that the Poovey Stack provides a technically credible pathway from optical switching to optical computing. The complete Boolean logic family is implementable. The von Neumann architecture can be replicated optically, or departed from in ways that exploit photonic parallelism. The AI workloads driving the infrastructure crisis map directly to operations that optical processors perform with extraordinary efficiency. The pieces fit together in ways that previous photonic computing approaches could not achieve. Whether those pieces can be assembled into working systems—whether the integration challenges can be solved, the manufacturing yields achieved, the commercial products delivered—remains to be demonstrated. The physics is proven. *The engineering is underway.*

But for a world facing an infrastructure crisis driven by the thermal consequences of electronic computation, the possibility of a different approach—one grounded in light rather than electrons, in femtoseconds rather than nanoseconds, in microwatts rather than watts—deserves the serious attention that this book attempts to provide.

CHAPTER 11

Building the Machine

The semiconductor industry has a saying: **the lab is not the fab.**

A technology that works in a research environment—with hand-tuned equipment, expert operators, unlimited time per device, and no requirement for profitability—may fail completely when transferred to manufacturing conditions. The history of advanced materials is littered with laboratory demonstrations that never became products: superconducting electronics that required helium cooling no factory could maintain, compound semiconductors that demanded precursor gases too toxic for industrial use, nanoscale structures that formed perfectly once but could not be reproduced reliably at the thousand-wafer-per-week pace that commercial production requires.

The previous two chapters described what the Poovey Stack can do: femtosecond switching, complete Boolean logic, efficient mapping to AI workloads. This chapter addresses whether it can be built—not once, in a laboratory, by inventors who understand every nuance of the technology, but repeatedly, in factories, by equipment and processes that must work the same way every time.

The question is not whether photonic logic is possible; Chapter 10 established that it is. The question is whether photonic logic is manufacturable at the scale and cost that commercial deployment requires. The answer depends on a strategic choice that distinguishes True Photonic's approach from previous photonic computing efforts. That choice is called Relaxed Lithography, and understanding it requires understanding why semiconductor manufacturing has become so expensive, so concentrated, and so difficult for new technologies to enter.

• • •

Modern electronic chips are manufactured through photolithography—a process that uses light to transfer patterns onto silicon wafers. The principle is straightforward: coat a wafer with light-sensitive material, shine light through a patterned mask, develop the exposed areas to create a pattern, then etch or deposit material through that pattern. Repeat dozens of times with different masks, and the accumulated layers form transistors, interconnects, and all the structures that constitute an integrated circuit.

The challenge is resolution. The smallest features that photolithography can produce are limited by the wavelength of light used—shorter wavelengths enable smaller features. For decades, the industry used deep ultraviolet light at 193 nanometers, achieving feature sizes down to approximately 40 nanometers through increasingly elaborate optical tricks: immersion lithography that placed liquid between the lens and wafer, multiple patterning that exposed each layer several times with offset masks, computational lithography that pre-distorted masks to compensate for optical aberrations.

By the early 2010s, these tricks had reached their limits. The industry needed a fundamentally shorter wavelength to continue shrinking features. The solution was extreme ultraviolet lithography—EUV—using light at 13.5 nanometers, more than fourteen times shorter than the previous generation. EUV enables features below 10 nanometers, the dimensions required for the most advanced chips in production today. The cost of this transition has been staggering. An EUV lithography machine costs approximately 350 million dollars. *A single machine.* A modern semiconductor fabrication facility—a fab—requires multiple EUV machines along with hundreds of other tools, cleanroom infrastructure, and support systems. The total cost to build a leading-edge fab now exceeds twenty billion dollars.

Only three companies in the world—TSMC, Samsung, and Intel—operate fabs capable of producing chips at the most advanced nodes. Only one company in the world—ASML, based in the Netherlands—manufactures EUV lithography equipment. This concentration creates a bottleneck for any technology that requires leading-edge manufacturing. The capacity is limited, the cost is enormous, and the queue is long. TSMC's advanced nodes are booked years in advance by Apple, Nvidia, AMD, and other high-volume customers willing to pay premium prices. A startup with a novel technology—however promising—cannot simply purchase access to EUV manufacturing. The economics do not work at low volumes, and the capacity is not available regardless.

The concentration has only intensified as Moore's Law has slowed. When each process generation delivered forty percent improvement, companies invested in their own fabs to capture competitive advantage. Now that improvements come in single-digit percentages at costs measured in billions, most chip designers have become fabless—designing chips but outsourcing manufacturing to the few companies that can afford the capital investment.

This model works for established products with predictable volumes, but creates barriers for new technologies that cannot guarantee the demand foundries require to justify their attention. Previous photonic computing efforts largely accepted this constraint and attempted to work within it. They designed devices for leading-edge nodes, assuming that photonic processors would need the smallest possible features to achieve competitive density and performance. This assumption shaped their architectures, their manufacturing partnerships, and ultimately their fates. When EUV capacity proved unavailable or unaffordable, the technologies stalled—not because the physics failed, but because the manufacturing pathway was blocked.

• • •

Relaxed Lithography™ rejects the assumption that photonic computing requires leading-edge processes. The Poovey Stack is designed for fabrication at mature process nodes—90 to 180 nanometers—using equipment and techniques that the semiconductor industry perfected fifteen to twenty years ago. No EUV required. No immersion lithography required. No multiple patterning required. The technology uses the manufacturing infrastructure that exists abundantly, operates reliably, and costs a fraction of leading-edge alternatives.

The strategy seems counterintuitive. If smaller features enable faster, denser, more efficient electronics, why would a technology claiming superior performance deliberately choose larger features? The answer lies in the fundamental difference between electronic and photonic devices. Electronic transistors switch by controlling the flow of electrons through channels measured in nanometers. Smaller channels mean shorter distances for electrons to travel, faster switching, lower capacitance, reduced power consumption. The relentless drive to smaller features in electronics reflects the physics of electron transport: smaller genuinely is better, up to the quantum mechanical limits now being approached.

Photonic devices operate through different physics.

Light at the wavelengths used in the Poovey Stack—1,550 to 1,600 nanometers in the telecommunications infrared band—has wavelengths measured in micrometers, not nanometers. The waveguides that confine and route optical signals must be sized to match these wavelengths; structures smaller than the wavelength cannot guide light effectively. The switching nodes that perform logical operations interact with light through phenomena that occur over hundreds of nanometers. The relevant dimensions for photonic devices are inherently larger than the leading-edge dimensions of electronic devices—not because of manufacturing limitations, but because of optical physics.

This mismatch between electronic and photonic length scales creates an opportunity. A photonic device fabricated at a 90-nanometer process node is not compromised by the "large" feature size; the features are still far smaller than the optical wavelengths being manipulated. The device can achieve its full theoretical performance—femtosecond switching, microjoule energy consumption, complete logic functionality—without requiring the extreme resolution that electronic devices demand. *The manufacturing simplicity is a benefit without a corresponding cost.*

The Relaxed Lithography approach is itself protected intellectual property, documented in True Photonic's patent portfolio. The filings establish that the methodology—achieving high-performance photonic computing through deliberate use of mature fabrication nodes—represents a novel manufacturing strategy, not merely a cost-saving compromise.

The mature-node strategy also avoids the yield penalties that plague leading-edge manufacturing. At 3-nanometer nodes, defect densities are measured in defects per square centimeter; at 90-nanometer nodes, they are measured in defects per square meter—a thousand-fold difference that translates directly into manufacturing yield. A technology designed for mature nodes inherits the yield advantages that decades of process refinement have achieved, rather than struggling with the yield challenges that leading-edge processes continuously confront.

The economic implications are substantial. A 90-nanometer fab costs hundreds of millions of dollars to build, not tens of billions. Dozens of such fabs operate worldwide, many with available capacity as their original customers have migrated to smaller nodes. The equipment is proven, the processes are characterized, the yields are high, the costs are known. A technology designed for these nodes can access manufacturing capacity that leading-edge technologies cannot—not in years, but in months.

• • •

The manufacturing process for Poovey Stack devices begins with the substrate described in Chapter 9: sapphire wafers with graphene grown directly through microwave plasma-enhanced chemical vapor deposition.

This step occurs outside the traditional semiconductor fab, in equipment designed for the specific requirements of graphene synthesis. The MW-PECVD process deposits high-quality graphene in under three minutes per wafer—a timeframe compatible with high-volume production and a dramatic improvement over the hours-long processes that characterized earlier graphene manufacturing approaches.

The significance of the three-minute cycle time deserves emphasis. Semiconductor manufacturing economics depend critically on throughput—the number of wafers processed per hour, per day, per year. Equipment that processes one wafer per hour, however excellent its results, cannot support volume production. Equipment that processes twenty wafers per hour can.
The MW-PECVD process, developed by Astera Energy and adapted for True Photonic's requirements, achieves the throughput that commercial manufacturing demands without sacrificing the material quality that device performance requires.

The graphene-on-sapphire substrates then enter a standard semiconductor fabrication flow, modified to accommodate the different materials but using the same fundamental process categories that the industry has refined over decades. Thin film deposition adds the boron nitride isolation layers and the piezoelectric materials that enable switching. Photolithography patterns each layer according to the device design. Etching removes material from areas not protected by the photoresist. The sequence repeats, layer by layer, building the three-dimensional stack structure that the architecture requires. The layer count varies by device complexity. Simple switching elements require five to ten layers. Logic circuits incorporating multiple gates require more. Complex integrated systems—the photonic processors that would eventually replace electronic chips for AI workloads—could require one hundred fifty layers or more. The Poovey Stack architecture supports this range, with the layer-by-layer fabrication approach allowing inspection and validation at intermediate stages rather than only at final test.

Several aspects of the process differ from conventional silicon fabrication. The thermal budgets must accommodate graphene's properties—high temperatures that would be routine for silicon processing could damage the two-dimensional carbon lattice. The etch chemistries must selectively remove some materials while preserving others in a materials stack more diverse than typical silicon devices.

The alignment tolerances, while relaxed compared to leading-edge electronics, must still maintain the precision that optical structures require—misalignment of waveguide layers by even tens of nanometers could introduce unacceptable losses. The deposition processes for the various layers require careful optimization. Boron nitride must be deposited in its hexagonal crystalline form to achieve the desired electrical and thermal properties; amorphous or randomly oriented material would compromise device performance. The piezoelectric layers must achieve the crystalline orientation that maximizes electromechanical coupling. These requirements add process constraints but do not require exotic equipment—the deposition techniques are variants of processes used throughout the semiconductor industry, adapted for the specific materials the Poovey Stack employs.

These differences require process development—the systematic work of establishing recipes, characterizing variations, and qualifying the manufacturing sequence for production. True Photonic has completed this development for the initial device structures, working with foundry partners to demonstrate that the processes produce working devices with acceptable yields. The work is not complete; more complex structures require additional development, and the transition from engineering samples to volume production involves further qualification. But the fundamental compatibility of the Poovey Stack with semiconductor manufacturing has been established through actual fabrication, not merely theoretical analysis.

• • •

Manufacturing yield—the percentage of devices that work correctly—determines whether a technology is commercially viable. A process with fifty percent yield wastes half its materials, equipment time, and labor producing scrap. A process with ninety-five percent yield converts nearly all inputs into saleable products. The difference between these scenarios can be the difference between profit and bankruptcy. Yield depends on defect density: the number of flaws per unit area that cause device failure. Defects arise from particles in the cleanroom, impurities in materials, variations in process conditions, equipment malfunctions, and dozens of other sources that semiconductor fabs spend billions of dollars controlling. Mature processes at established nodes achieve defect densities low enough that large chips can be manufactured economically. Leading-edge processes, still being refined, typically suffer higher defect densities that improve over time as engineers identify and eliminate sources of variation.

The Poovey Stack benefits from the maturity of the processes it employs.

A 90-nanometer process in 2025 has been running in production for nearly two decades. The equipment is understood, the failure modes are characterized, the defect sources have been systematically eliminated. Yields at these nodes routinely exceed ninety percent for conventional devices. The introduction of new materials—graphene, boron nitride, the piezoelectric layers—creates new potential defect sources, but the baseline process stability provides a foundation that leading-edge manufacturing cannot offer. The layer-by-layer fabrication approach also enables yield management strategies unavailable in conventional chip manufacturing. Each layer can be inspected before subsequent layers are added. Defective regions can be identified early, before additional processing wastes resources on devices that will ultimately fail. In principle, defective sections could be repaired or bypassed through redundancy designed into the architecture—approaches that the vertical structure of the Poovey Stack accommodates more readily than the planar structure of conventional chips.

The economic model for photonic computing differs from electronics in another important respect: the value per device can be substantially higher. An electronic processor for AI applications—an Nvidia H100 GPU, for example— sells for approximately 25,000 to 40,000 dollars. If a photonic processor delivers equivalent computational capability at a fraction of the power consumption, the value proposition supports similar or higher pricing even at lower manufacturing volumes. The economics do not require the millions of units per year that justify leading-edge electronic fabs; they require thousands of units at prices that reflect the operational savings the technology delivers.

Consider the total cost of ownership for a data center operator. The purchase price of processors is significant but not dominant; electricity costs over a multi-year deployment often exceed hardware costs. A photonic processor priced at fifty thousand dollars that reduces electricity consumption by ninety percent could deliver payback in under a year for operators currently spending millions annually on power for AI computation. The economics favor photonic computing even at price points that would be untenable for consumer electronics—because the customers are not consumers, but enterprises whose operating costs dwarf their capital costs.

• • •

The path from current capabilities to commercial products follows a progression familiar from semiconductor history: device validation, circuit integration, subsystem demonstration, system qualification, and volume production. True Photonic has completed the first stage—individual switching devices have been fabricated and validated, including the independent testing at Technion described in Chapter 9. The subsequent stages involve progressively more complex integration, with each stage validating both the technology and the manufacturing processes before advancing. The immediate next step is circuit integration—combining multiple switching nodes into functional logic circuits that perform defined operations. A circuit containing tens of switches, implementing basic arithmetic or control functions, validates that the switches work together as designed—that the timing aligns, the signal levels are compatible, the thermal interactions are manageable. This stage is currently underway, with engineering samples in fabrication and testing.

The circuit integration stage reveals challenges invisible at the device level. Individual switches, tested in isolation, experience uniform optical conditions—controlled input power, defined wavelengths, characterized thermal environment. Switches integrated into circuits must operate amid the optical signals from neighboring switches, the thermal gradients from simultaneous switching events, and the accumulated losses from propagation through multiple waveguide segments. A switch that performs flawlessly in isolation may exhibit degraded performance in context. The circuit integration stage identifies these effects and develops mitigation strategies before they compound at larger scales.

Subsystem demonstration follows: larger aggregations of circuits that perform useful computational tasks. A photonic matrix multiplier, for example, would combine thousands of switching nodes in the crossbar configuration described in Chapter 10, accepting input vectors and producing output vectors that represent the matrix product. This stage validates not only the technology but its application—demonstrating that photonic computation actually delivers the efficiency advantages that the physics predicts.

The subsystem stage also establishes the design methodology for larger systems. What software tools support circuit design at this scale? What simulation models predict performance accurately? What test procedures verify functionality efficiently? These questions have well-established answers for electronic design; photonic design requires developing equivalent capabilities.

The subsystem stage produces not only working hardware but the design infrastructure that subsequent development requires. System qualification integrates photonic subsystems with the electronic infrastructure required for practical deployment: input/output interfaces, control systems, packaging, and cooling. Even photonic processors that generate minimal heat require thermal management for stability; even optical systems require electronic interfaces for communication with the broader computing environment. This stage produces complete products that can be evaluated by customers in realistic operating conditions.

The packaging challenge deserves particular attention. Electronic chips are packaged in standardized formats—ball grid arrays, land grid arrays, and similar configurations that connect to printed circuit boards through well-established interfaces. Photonic devices require optical connections as well as electrical ones: fiber-optic interfaces that couple light into and out of the chip, alignment mechanisms that maintain optical pathways, protective enclosures that shield sensitive structures from contamination. These requirements add complexity but are not unprecedented; the telecommunications industry has developed photonic packaging for decades, and that experience informs the approaches True Photonic is adapting for computing applications.

Volume production scales the qualified processes to manufacturing rates that support commercial deployment. The transition from engineering volumes—tens or hundreds of devices—to production volumes—thousands or tens of thousands—requires process optimization for throughput, supply chain development for materials and components, quality systems for consistency, and all the infrastructure that distinguishes a laboratory from a factory. The volume production stage also establishes the supply chain resilience that commercial customers require. A technology dependent on single-source materials or sole-provider processes creates business risk that enterprise customers will not accept. True Photonic's strategy emphasizes multiple qualified suppliers for critical materials, multiple qualified foundries for device fabrication, and standard interfaces that avoid lock-in to particular equipment or process variants.

The Relaxed Lithography approach supports this strategy by accessing manufacturing capacity that exists across multiple geographies and multiple providers.

True Photonic's timeline projects eighteen to twenty-four months from current status to initial commercial products. This estimate reflects the device validation already completed, the circuit integration currently underway, and realistic assessments of the engineering work remaining. The timeline could extend if unexpected challenges emerge during integration; it could compress if the work proceeds without significant obstacles. What the timeline does not include is the decade-long development cycles that characterized previous photonic computing efforts—cycles driven largely by manufacturing challenges that Relaxed Lithography avoids.

• • •

The geographic distribution of semiconductor manufacturing has become a matter of national security concern. Leading-edge fabs are concentrated in Taiwan and South Korea, with limited capacity in the United States, Europe, and Japan. This concentration creates supply chain vulnerabilities that governments worldwide are now attempting to address through industrial policy, subsidies, and strategic investments.

The CHIPS Act in the United States, the European Chips Act, and similar initiatives in Japan and elsewhere all aim to reshore semiconductor manufacturing capacity. These initiatives focus primarily on leading-edge manufacturing—the EUV-based processes that produce the most advanced chips. The assumption is that leading-edge capacity is strategically critical, and that mature-node capacity is adequately distributed and less significant. For conventional electronics, this assumption is reasonable.

For photonic computing, it creates an opportunity. Mature-node manufacturing capacity exists abundantly in the United States and allied nations. Fabs that once produced leading-edge chips have transitioned to mature nodes as their equipment aged; new fabs serving automotive, industrial, and analog markets operate at these nodes by design. A technology that requires 90 to 180 nanometer processes—not 3 to 7 nanometer processes—can be manufactured domestically without the multi-billion-dollar investments that leading-edge reshoring requires. The strategic implications extend beyond simple supply chain security. A domestically manufactured photonic computing technology could reduce dependence on the concentrated leading-edge capacity that currently constrains AI development. Data centers powered by photonic processors would not compete for the limited TSMC capacity that every AI company currently requires. The bottleneck that Chapter 2 described—the shortage of advanced chips that limits AI deployment—could be bypassed rather than merely alleviated.

The policy implications are significant. Governments investing billions to reshore leading-edge semiconductor manufacturing face timelines measured in decades and outcomes measured in uncertainty. A photonic computing technology manufacturable at mature nodes could deliver strategic capability sooner, at lower cost, using infrastructure that already exists. This is not an argument against leading-edge investments—electronic computing will continue to require advanced manufacturing for the foreseeable future—but rather an observation that photonic computing offers a complementary pathway with different risk and timeline characteristics.

This is not to suggest that photonic computing will immediately replace electronic computing across all applications. Electronic processors will continue to dominate general-purpose computing, mobile devices, and applications where the existing infrastructure represents sufficient capability. But for the specific workloads driving the infrastructure crisis—the AI inference that consumes megawatts per data center, the training runs that require dedicated power plants—photonic alternatives manufactured outside the leading-edge bottleneck could provide both performance and strategic advantages.

• • •

The honest assessment of manufacturing readiness must distinguish between what has been demonstrated and what remains to be proven.

True Photonic has fabricated working devices at commercial foundries using the Relaxed Lithography approach. The switches work. The processes are compatible with standard semiconductor manufacturing. The substrate technology produces material suitable for device fabrication. These are not projections; **they are accomplishments.**

What has not been demonstrated is volume production of complex integrated systems. The engineering from validated switches to validated circuits is underway but incomplete. The further engineering from circuits to subsystems to products remains ahead. Each transition introduces challenges that cannot be fully anticipated until encountered. The eighteen-month timeline represents management's best estimate, not a guaranteed delivery date.

The manufacturing yields that will be achieved in production are unknown. Process development has established that working devices can be fabricated, but the yield percentages that determine commercial viability will emerge only from production experience. True Photonic's projections assume yields consistent with mature-node manufacturing of novel device structures—reasonable but unvalidated assumptions that production will eventually confirm or refute.

The cost structures that will characterize commercial products remain uncertain. Device costs depend on yields, throughput, materials costs, and manufacturing overhead—all of which will be refined through production experience. The pricing assumptions in True Photonic's business model are based on engineering estimates and market analysis, not production accounting.

What can be said with confidence is that the manufacturing pathway is realistic. Unlike previous photonic computing efforts that required exotic materials, cryogenic temperatures, or processes available only in specialized research facilities, the Poovey Stack can be built using existing semiconductor infrastructure at accessible cost points. The pathway is not guaranteed to succeed, but it is not blocked by fundamental manufacturing barriers.

The lab may not be the fab, but the gap between them has been deliberately minimized.

• • •

Part III of this book has examined the architecture of photonic computing: the physical structures, the logical capabilities, and the manufacturing pathways that could enable a new approach to computation. The Poovey Stack provides validated switching at femtosecond speeds. The logic family achieves functional completeness through mechanisms compatible with commercial production. The Relaxed Lithography strategy accesses manufacturing capacity that leading-edge electronic technologies cannot reach.

Part IV turns from how the technology works to what it could mean—for the infrastructure crisis described in Part I, for the communities affected by data center expansion, for the industries that depend on continued growth in computational capability. The examination will be prospective rather than retrospective, exploring possibilities rather than documenting accomplishments.

The uncertainties acknowledged throughout Part III apply with greater force to the projections that follow.

Chapter 12 examines Clean Compute: the possibility of data centers that operate without the resource consumption that has made them controversial. If photonic processors deliver the efficiency the physics predicts, what does that mean for facility design, for site selection, for community relations? The infrastructure of computing could be fundamentally reconfigured.

Chapter 13 examines the new stack: the applications and industries that photonic computing could enable. From medical diagnostics to financial modeling to autonomous systems, computational capability has historically unlocked new possibilities. Photonic computing at photonic efficiency could unlock possibilities currently foreclosed by power and thermal constraints.

Chapter 14 examines the light ahead: the trajectory from current status to widespread deployment, the obstacles that remain, and the reasons for believing that 2026 might indeed become the year of light. The conclusion will be neither triumphalist nor defeatist, but realistic—acknowledging both the genuine promise and the genuine uncertainty that characterize any emerging technology.

The architecture has been described. The manufacturing has been explained. What remains is to explore what could be built—**and why it matters.**

Computation that does not require rivers of water, forests of cooling towers, and the output of power plants dedicated to answering questions.

PART IV

THE FUTURE

The year 2026 may or may not become the year of light. What happens next depends on execution

CHAPTER 12

Clean Compute

The data center that dominates the public imagination is a monument to industrial necessity: a windowless fortress covering acres of land, surrounded by cooling towers that exhale clouds of steam into the atmosphere, humming with the constant drone of fans and chillers, drawing power through transmission lines thick as a man's arm from substations that could serve a small city.

This image is accurate because it reflects the physical requirements of electronic computation at scale. The machines generate heat. The heat must be removed. The removal requires infrastructure that cannot be hidden. The communities that host these facilities experience them as industrial installations—because that is what they are. The tax revenue may be welcome, the jobs may be appreciated, but the facility itself is not something anyone would choose to live beside. The noise, the traffic during construction, the visual intrusion of industrial architecture on landscapes that were recently farms or forests—these are costs that communities accept in exchange for economic benefits, or resist when the benefits seem insufficient.

Part I of this book documented the backlash that has emerged as these costs have become clearer: the sixty-four billion dollars in blocked or delayed projects, the activist groups organizing in twenty-eight states, the local officials losing elections over data center approvals. The opposition crosses political lines and geographic boundaries. In Warrenton, Virginia, five hundred residents packed a town council meeting to oppose Amazon. In Chandler, Arizona, the city council voted unanimously against a facility despite lobbying from a former United States Senator. The backlash is not irrational; it reflects a genuine assessment that the costs and benefits of current data center development are not fairly distributed.

But what if the data center itself could change? What if the dramatic power reductions described in Part III translated into facilities that looked fundamentally different from the industrial installations that communities have learned to oppose? This chapter examines that possibility—not as speculation about distant futures, but as an exploration of what photonic computing's efficiency could mean for data center design, site selection, and community relations if the technology delivers what the physics predicts. The physics of the transformation is straightforward, even if the implications are far-reaching. A data center's physical footprint is determined primarily by its cooling requirements. The servers themselves are relatively compact—racks of circuit boards and processors that pack enormous computational density into modest volumes. But those processors generate heat, and heat removal dominates facility design.

Consider a hypothetical one-hundred-megawatt data center, representative of the hyperscale facilities operated by major cloud providers. One hundred megawatts of electrical power, converted almost entirely to heat by the electronic processors, must be continuously removed from the building. The cooling infrastructure required for this task is substantial: massive chillers that could fill a basketball court, cooling towers rising forty or fifty feet into the sky, pumping systems that circulate hundreds of thousands of gallons of water daily, air handling units the size of shipping containers pushing millions of cubic feet of air per hour. The cooling plant can occupy as much floor space as the computing equipment it serves. The cooling towers alone may cover an acre or more.

The numbers are not abstractions. Google's facility in Council Bluffs, Iowa—one of the largest in the world—consumes approximately one billion gallons of water annually. That is enough water to supply the residential needs of a city of fifty thousand people. Microsoft's data centers in Arizona draw water from the Sonoran Desert's already-stressed aquifer systems. Northern Virginia's data center cluster consumed nearly two billion gallons in 2023, a sixty-three percent increase from just four years earlier. These facilities are not merely using water; they are consuming it, sending it into the atmosphere through evaporative cooling where it disperses beyond recovery.

Now apply the efficiency gains that photonic computing offers. *The same computational output—the same queries answered, the same models trained, the same data processed*—requires ten megawatts instead of one hundred. Ten megawatts of heat to remove instead of one hundred. The cooling infrastructure scales accordingly, not quite linearly but close enough for planning purposes. The cooling towers that dominated the site become unnecessary. The chiller plant that occupied a significant portion of the building shrinks to a fraction of its former size. The pumping systems, the air handlers, the elaborate ductwork—all reduce in proportion to the reduced thermal load.

The transformation is not merely quantitative. Below certain thermal thresholds, the nature of cooling changes entirely. A facility generating one hundred megawatts of heat requires active cooling—mechanical systems that consume significant energy themselves, typically adding fifteen to thirty percent to the facility's total power draw. A facility generating ten megawatts can potentially be cooled passively, or with minimal mechanical assistance. The heat can be dissipated through conventional building HVAC systems, through carefully designed natural ventilation, through thermal mass and building envelope design that would be utterly inadequate for electronic computing.

The metric that data center operators use to measure cooling efficiency is called Power Usage Effectiveness, or PUE. A PUE of 2.0 means that for every watt powering computing equipment, another watt powers cooling and other overhead. A PUE of 1.5 means that overhead consumes half as much as computation. The best hyperscale facilities achieve PUEs around 1.1 or 1.2 through extraordinary engineering investment. A photonic facility with dramatically less heat to remove could achieve PUEs approaching 1.0—essentially no cooling overhead—because the heat load would fall below the threshold requiring dedicated cooling infrastructure. The water consumption that has made data centers controversial in drought-prone regions follows the same logic. Evaporative cooling—the cooling towers that consume millions of gallons annually—exists because it is the most efficient way to reject large amounts of heat. When the heat load drops substantially, evaporative cooling may no longer be necessary. Closed-loop systems that reject heat without consuming water become practical.

Air-cooled systems that would be inadequate for electronic data centers can handle photonic thermal loads. *The facility that once drank a river needs only modest municipal water service.*

• • •

The visual transformation would be equally dramatic. The cooling towers that announce a data center's presence from miles away—those massive concrete or steel structures venting plumes of water vapor into the sky—would simply not exist. The backup generators that line facility perimeters, ready to maintain operations during power outages, could be smaller or fewer. The electrical substations that can cover acres would shrink proportionally. The facility would look like what it is: a building containing computing equipment. **Not a power plant. Not a chemical factory.** *A building.*

The architectural possibilities that emerge from this transformation are significant. Current data centers are designed around cooling constraints: floor layouts optimized for hot-aisle and cold-aisle airflow patterns, ceiling heights of fifteen or twenty feet to accommodate air handling requirements, building orientations chosen for cooling tower placement and prevailing wind patterns. The buildings are utilitarian because utility is all that matters when thermal management dominates design. Remove the thermal constraints, and data centers could be designed like other commercial buildings—for efficiency of space utilization, for aesthetic compatibility with surrounding development, for integration into urban fabric rather than isolation from it. The ceiling heights could drop to standard commercial dimensions. The floor plates could be configured for human occupancy rather than optimized airflow. Windows become possible when the building no longer needs to be a hermetically sealed thermal management system. The architecture could express something other than industrial necessity.

The footprint reduction compounds these possibilities. A photonic data center delivering equivalent computational output to a current hyperscale facility might require one-tenth the land area—not because the computing equipment is smaller, but because the cooling infrastructure that dominates current facilities would largely disappear. A facility that once required forty acres might fit on four. A facility that demanded a rural site because of its scale and visual impact might be appropriate for suburban or even urban locations.

This is not to suggest that every data center could become an urban high-rise. Large-scale computing will continue to require significant infrastructure: robust electrical service with multiple redundant feeds, diverse network connectivity from multiple carriers, physical security appropriate for critical infrastructure, and access for maintenance and equipment replacement. But the constraints that have pushed data centers to remote locations—primarily the need for space and resources to support massive cooling operations—would be fundamentally relaxed. The site selection calculus would change.

• • •

SkyLight Holdings Inc. was created to pursue exactly this opportunity—but with a focus that goes beyond building new facilities. The insight driving SkyLight is that photonic computing does not merely enable a new kind of data center; it unlocks an entirely new category of real estate for computational use.

Specifically, it transforms stranded commercial towers from depreciating liabilities into appreciating infrastructure assets.

The commercial real estate crisis that followed the pandemic has left cities worldwide with a specific and growing problem: office towers that cannot sustain their original purpose. Remote work has permanently reduced demand for traditional office space. Occupancy rates in major markets remain twenty to thirty percent below pre-pandemic levels, with no clear path to recovery. The buildings exist—substantial structures with robust electrical service, elevator systems, floor loading designed for commercial use—but the tenants who once filled them are not coming back. The obvious solution, residential conversion, has proven economically challenging in most cases. Office floor plates are typically too deep for residential use, with interior spaces too far from windows for livable apartments. The plumbing infrastructure is wrong—offices need restrooms clustered around cores, while residences need kitchens and bathrooms distributed throughout. The conversion costs frequently exceed the value of the resulting residential units, particularly in markets where residential rents cannot support the investment required.

The result is a class of stranded assets measured in the hundreds of billions of dollars globally. Real estate investment trusts hold these buildings on their balance sheets at values that bear diminishing relationship to economic reality. Pension funds that invested in commercial real estate face write-downs that affect millions of beneficiaries. Municipal tax bases built on commercial property assessments are eroding as values decline. The problem is structural, not cyclical, and traditional real estate solutions have not solved it.

Photonic computing offers a different kind of conversion—one that works with the characteristics of office towers rather than against them. Deep floor plates that frustrate residential conversion are ideal for computing equipment, which does not need window access. Robust electrical service designed for commercial loads can power computational infrastructure. Elevator systems built for office worker traffic can handle equipment installation and maintenance. The buildings were designed for exactly the kind of use that photonic data centers require, minus the thermal constraints that have historically made such use impossible. The thermal constraints are the key. A traditional data center cannot occupy an office tower because the cooling requirements are incompatible with building systems designed for human comfort.

An office building's HVAC system might handle ten or twenty watts per square foot of heat load; a data center generates one hundred watts per square foot or more. The mismatch is an order of magnitude. No amount of retrofit engineering can make an office tower's cooling infrastructure adequate for electronic computing at scale. But photonic computing changes the equation. A facility generating dramatically less heat falls within the range that commercial building systems can manage. The building's existing HVAC, perhaps modestly upgraded, can handle the load. The transformation from stranded office tower to operational data center becomes a matter of installing computing equipment and network connectivity, not reconstructing the building's mechanical systems from scratch.

• • •

The strategic implications of this insight extend beyond individual buildings to the structure of the technology industry's support. Data centers have historically created opposition: communities that resist their construction, ratepayers who bear higher electricity costs, environmental groups that challenge their resource consumption. The political economy of data center development is one of concentrated benefits—to operators and their customers—and distributed costs borne by everyone else. Tower conversion creates a different political economy. Real estate owners holding stranded assets become advocates for the technology that makes those assets valuable again. Institutional investors—the pension funds, insurance companies, and sovereign wealth funds that hold commercial real estate—develop financial interest in photonic computing's success. Municipal governments facing eroding tax bases see a path to restoring property values without the burdens that traditional data centers impose. The stakeholder alignment shifts from opposition to support.

Consider the magnitude of the aligned interests. The commercial real estate market in the United States alone represents trillions of dollars in asset value, with office properties comprising a substantial share. The institutional owners of these assets—Blackstone, Brookfield, pension systems like CalPERS and the Teachers' Retirement System of Texas—manage capital measured in hundreds of billions. If photonic computing can transform even a fraction of stranded office inventory into productive infrastructure, these institutions become natural advocates for the technology's adoption and success. SkyLight Holdings, as a subsidiary of True Photonic—eighty percent owned by the parent company—is positioned to capture this opportunity while creating these aligned stakeholders.

The corporate structure reflects a strategic assessment: True Photonic's core business is developing the photonic computing technology itself, while building and operating data centers requires different capabilities, different capital structures, and different relationships. SkyLight was established to pursue the infrastructure opportunity, to build the partnerships with real estate owners that tower conversion requires, and to demonstrate that photonic computing creates value beyond the technology itself.

The name carries deliberate significance. "SkyLight" evokes both the light-based computation that enables the facilities and the architectural transformation that becomes possible—towers with their original windows and natural light preserved, not hermetically sealed industrial installations. But the name also signals the strategic focus: not greenfield development on rural sites, but vertical deployment in the towers that define urban skylines worldwide. The towers serve purposes beyond computation alone. Their height and urban locations make them natural nodes in communications networks. True Photonic's patent portfolio includes innovations in optical signal distribution that could transform converted towers into infrastructure serving multiple functions: computation, certainly, but also network connectivity that extends the reach of photonic systems beyond the buildings themselves. The details of this network architecture—including the use of existing structures like billboards as signal repeaters—belong to Chapter 14's discussion of the broader infrastructure vision. Here, the essential point is that tower conversion is not merely about repurposing real estate; it is about creating an integrated infrastructure that compounds value across multiple dimensions.

• • •

The urban deployment that tower conversion enables deserves examination on its own terms. Current hyperscale facilities are located where they are—in rural Iowa, in suburban Virginia, in the Arizona desert—because urban locations cannot accommodate their requirements. The land costs would be prohibitive for facilities covering twenty or forty acres. The power requirements would strain urban grids already serving dense populations. The cooling infrastructure would be impossible to site amid surrounding development.

Photonic computing in converted towers inverts this logic entirely. The buildings already exist; there are no land costs for new construction.

The electrical service is already in place, designed for commercial loads that approximate what photonic computing requires. The cooling can be handled by building systems designed for human occupancy. The infrastructure barriers that have excluded computation from urban cores dissolve.

The advantages of urban deployment are substantial for certain applications. Latency—the delay between a user's request and the system's response—decreases with proximity. Light travels through fiber optic cable at roughly two hundred thousand kilometers per second, but even at light speed, distance introduces measurable delay. A data center in Manhattan serves New York users faster than a data center in Iowa by perhaps twenty or thirty milliseconds. For most web browsing, that difference is imperceptible. For financial trading, it is the difference between executing a trade and missing it. For real-time collaboration, it is the difference between natural conversation and awkward delay. For autonomous vehicles receiving updates from cloud systems, it could be the difference between safe operation and dangerous lag.

Current edge computing strategies attempt to address this by deploying smaller facilities closer to users. Companies like Equinix and Digital Realty operate urban data centers in major metropolitan areas, accepting higher costs and smaller scales to achieve proximity benefits. But edge facilities face the same thermal constraints as hyperscale centers, merely at smaller scale. They still require cooling, still generate noise, still present siting challenges. A five-megawatt edge facility needs proportionally the same cooling infrastructure as a fifty-megawatt hyperscale center. The heat per server is identical regardless of facility size.

Photonic edge facilities in converted towers would face these challenges in attenuated form—potentially mild enough to allow deployment on multiple floors of a single building, or across a portfolio of buildings in a metropolitan area. The facility would be invisible to the building's other occupants if any remain—no different in impact from a large IT installation or trading floor server room. The computation that currently happens in distant data centers could happen in the same building where users work, or in the tower across the street, or distributed across dozens of towers throughout a city. The community relations transformation follows from these changes in physical impact and stakeholder alignment. The data center opposition documented in Part I emerges from a straightforward assessment: the facilities consume resources, create nuisances, and provide limited local benefit while enriching distant corporations. Photonic computing in converted towers changes every element of this equation.

The resource consumption drops dramatically. The nuisances—noise, visual impact, water consumption—largely disappear when cooling requirements fall below the threshold requiring industrial infrastructure. The local benefit expands when the facility preserves and enhances an existing building rather than replacing productive land with industrial installation. The enrichment flows in part to local stakeholders—the building owners, the municipal tax base, the workers who operate and maintain the facility—rather than exclusively to distant technology companies.

SkyLight's approach to community benefit goes beyond merely eliminating nuisances. Even a facility with substantially reduced heat generation produces some thermal output, and that output can be captured rather than wasted. District heating systems in Scandinavian countries already use data center waste heat to warm nearby buildings; a SkyLight facility could provide similar benefit to surrounding development, converting computational byproduct into community amenity. The buildings' existing infrastructure can be shared: the electrical service that powers computation can provide backup power to neighboring buildings, the network connectivity can extend to local institutions, the facility can include community-accessible spaces that transform a formerly vacant tower into a neighborhood asset. The economic model that makes these provisions possible reflects both photonic computing's efficiency and the tower conversion strategy's capital efficiency. Traditional data centers require billions in new construction; tower conversion repurposes existing assets at a fraction of the cost. Traditional data centers pay premium prices for rural land with adequate power and water; tower conversion uses urban buildings whose values have collapsed precisely because traditional uses no longer work. The savings compound: lower construction costs, lower land costs, lower operating costs from reduced power consumption. The margin created by these savings can fund community benefit that traditional data center economics cannot accommodate.

• • •

The honest acknowledgment that has characterized this book applies with particular force to the projections in this chapter. Nothing described here has been demonstrated at scale. The power reductions have been measured in device testing, not facility operation. The tower conversion model has been analyzed, not implemented. The stakeholder alignment has been theorized, not validated through actual transactions. The gap between device performance and facility performance is significant.

A photonic processor that achieves substantial power reduction in laboratory testing must maintain that efficiency when deployed in production systems, integrated with electronic components for input and output, operating continuously under varying loads for months and years. System integration introduces losses and overhead that device measurements do not capture. Engineering estimates suggest that system-level efficiency representing seventy to eighty percent power reduction—rather than the higher figures at device level—is realistic for initial deployments. This reduced figure still represents a transformational improvement, still enables the architectural and operational changes described in this chapter. But it is a more conservative projection than the device specifications might suggest.

The tower conversion model involves assumptions about building suitability that will vary case by case. Not every stranded office tower will be appropriate for computational use. Some buildings have electrical service inadequate for the load. Some have structural limitations that prevent equipment installation. Some have ownership structures or existing commitments that complicate conversion. The model describes a category of opportunity, not a universal solution applicable to every distressed commercial property.

The stakeholder alignment depends on demonstrating value before it can generate support. Real estate owners will not advocate for photonic computing based on theoretical analyses; they will advocate based on successful conversions that restore value to their portfolios. Institutional investors will not commit capital based on projections; they will commit based on operating facilities that generate returns. The political support that aligned stakeholders could provide remains potential until the first conversions demonstrate that the model works.

True Photonic and SkyLight understand this requirement and are acting accordingly. Prototype site development is being planned in Georgia and New Orleans, working toward the first operational demonstrations of the tower conversion model. The specific properties remain under evaluation, but the commitment to building proof points—not merely publishing analyses—reflects recognition that the model must be demonstrated before it can be scaled.

The approach is deliberately vertically integrated. True Photonic is building its own photonic computing stack for initial deployments rather than depending on external partners for critical components. SkyLight is developing its own facilities rather than licensing the model to third-party operators. This vertical integration sacrifices speed for control: by owning the technology, the manufacturing relationships, and the initial facilities, the companies can ensure that first deployments succeed on their own terms rather than failing due to execution problems outside their control. The first conversions must work. Everything that follows depends on them working. The strategy reflects that reality.

The regulatory environment for tower conversion is untested. Zoning codes in most jurisdictions do not contemplate data centers in office buildings. Building codes may require modifications to address the specific characteristics of computational equipment in spaces designed for human occupancy.

Permitting processes may introduce delays as officials encounter unfamiliar applications. These regulatory uncertainties add timeline risk to the tower conversion strategy. These uncertainties are not reasons for pessimism; they are reasons for caution and for staged implementation that validates assumptions before committing to scale. The technical foundation for Clean Compute is sound. The tower conversion thesis addresses a real problem with a logical solution. The stakeholder alignment creates genuine shared interest. But the translation from concept to deployment involves countless details that cannot be fully anticipated, and some of those details will prove more challenging than current projections assume.

• • •

The infrastructure crisis described in Part I of this book is not a problem that will solve itself. The demand for computation continues to grow at rates that strain even optimistic projections. The constraints on power, water, and community acceptance continue to tighten as existing resources are exhausted and new development faces organized opposition. The gap between what current technology requires and what sustainable development permits continues to widen.

Something must change, and the question is whether photonic computing can be that change. Clean Compute represents one vision of what that change could look like. Computation that operates in converted urban towers rather than purpose-built rural installations. Facilities that generate wealth for real estate owners rather than opposition from neighboring communities. Infrastructure that supports the continued growth of artificial intelligence while solving rather than creating problems for the cities where most people live.

The vision is ambitious, but it is not fantasy. It follows directly from the technical specifications validated in Part III of this book. If photonic processors achieve the efficiency that physics predicts and device testing has demonstrated, then the tower conversion model becomes economically and physically viable. The transformation is not speculative; it is consequential. The question is not whether these characteristics are possible but whether the technology can be deployed at the scale required to address both the infrastructure crisis and the commercial real estate crisis simultaneously.

SkyLight Holdings exists to answer that question. Its conversions will either demonstrate that Clean Compute in urban towers is practical—that photonic computing can restore value to stranded assets while delivering computational infrastructure—or reveal obstacles that require further engineering and alternative approaches. Either outcome advances understanding. Either outcome contributes to addressing the crises that motivated this book. The following chapters complete the examination of photonic computing's potential. Chapter 13 looks beyond infrastructure to the applications and industries that photonic efficiency could enable—the new capabilities that become possible when computational power is no longer constrained by thermal limits. Chapter 14 examines the path forward: the timeline for deployment, the broader network infrastructure that converted towers could support, and the reasons for believing that the next few years will determine whether photonic computing achieves its promise or joins the long list of technologies that were always almost ready.

The infrastructure is the foundation. Without facilities to deploy photonic processors, the technology remains academic—interesting to researchers but irrelevant to the crises it purports to address. Clean Compute is the bridge between laboratory achievement and practical impact—the means by which theoretical efficiency becomes measured benefit.

The towers are waiting. The technology is advancing. **What happens next depends on whether the bridge can be built.**

CHAPTER 13

The Next Stack

The previous chapters examined photonic computing as infrastructure—processors, data centers, the physical plant of computation. This chapter examines what becomes possible when that infrastructure exists. The applications are not hypothetical extensions of current technology; they are capabilities that electronic computing cannot achieve regardless of how much power it consumes or how many chips are deployed. They represent new categories of what computation can do.

The Poovey Stack is not merely a faster processor. It is a platform—a foundation on which applications can be built that exploit photonic computing's specific characteristics: *femtosecond switching speeds, negligible heat generation, extreme power efficiency*, and the *ability to operate in environments where electronic processors cannot survive.*

Each characteristic enables applications that were previously impossible, impractical, or prohibitively expensive. ***The platform creates a surface for innovation that extends far beyond data center efficiency.***

True Photonic's corporate structure reflects this platform logic. The parent company develops and licenses the core technology. SkyLight Holdings, as described in Chapter 12, deploys infrastructure. True Photonic Ventures creates and funds companies that build applications on the photonic platform—vertical-specific enterprises that take the technology into markets where domain expertise matters as much as technological capability. The ventures become tenants in SkyLight facilities, creating an ecosystem where infrastructure and applications reinforce each other.

• • •

I did not set out to reinvent money. I did not set out to rebuild the internet. I did not set out to transform national defense or revolutionize medicine.
I set out to solve a problem in computation, and the solutions kept pointing toward transformations I had never anticipated.

The pattern became clear as I learned what the Poovey Stack could actually do. Every time Del or Mark explained another characteristic of photonic switching—the power efficiency, the thermal profile, the speed—my mind raced to applications. Not incremental improvements to existing products, but wholesale replacements of systems the world assumes are fixed.

I started filing patents. Not just for the core technology—the team was handling much of that—but for every ***application*** I could be made to understand. *Blockchain infrastructure. Optical networking. Solar energy conversion. Cardiac monitoring. Robotic actuation. Defense systems.* Each vertical where photonic computing could replace electronic systems, I wanted True Photonic to own the intellectual property before the world understood what was coming.

Fifteen provisional patents filed between November 2025 and January 2026.
Supplementals to the the Poovey Stack itself.
A Photonic Muscle system for robotic actuation.
The graphene substrate platform.
TCP/IP-L networking (internet in light)
Photonic-graphene solar cells.
Relaxed Lithography manufacturing.
Storage-in-Light memory.
CardioLT wearable monitoring.
PulseLT implantable devices.
AI inference engines covering photonic tensor cores, transformers, and mixture-of-experts architectures.
Transaction verification for micropayment processing.
Cryptographic hash computation.
Outcome-verifiable multi-tenant platforms with deterministic SLA enforcement.

Each patent represents a vertical where we intend to be first—***not just first to market***, but first to own the intellectual framework that defines the market.

The window will not stay open forever. Once we open the first Clean Compute Center, once the efficiency numbers become public, once the industry realizes that photonic computing actually works—there will be a mad scramble.

Every major technology company, every chip manufacturer, every vertically integrated player will race to develop or acquire photonic capability. The patents we file now, for applications the market has not yet imagined, become the toll roads of that future.

Our internal tagline captures the scope of the ambition: ***The Next Generation of Everything.***

It is not hyperbole.

Every device that currently runs on electrons could, in principle, run on photons—**and run better**. Every system that generates heat could run cool. Every network that merely transports data could also compute during transit. Every financial transaction that takes minutes could settle in milliseconds. Every medical device that lasts years could last decades. The platform is horizontal; the applications are limitless.

This chapter surveys four domain examples where the next stack enables transformational applications: a new architecture for trust and value exchange, a new model for the internet itself, sovereign infrastructure for national security, and—at the most intimate scale—devices that operate inside the human body. The examination is necessarily forward-looking; these applications are in development, not deployment. But the physics that enables them is validated, and the path from current capabilities to these applications is engineering rather than discovery.

• • •

The Light-Switch Ledger

Bitcoin *mined coal.* Ethereum *mined electricity.* ***We mint with photons.***

That is not a slogan. It is a description of what becomes possible when the fundamental unit of computation changes from an electronic transistor to a light-actuated light-switch. The blockchain revolution promised to transform finance, governance, and trust itself—but it ran into a wall built of physics. The energy required to secure distributed ledgers grew until Bitcoin alone consumed more electricity than Thailand. The heat generated by mining operations required cooling infrastructure that drained aquifers and strained power grids. The carbon footprint expanded until cryptocurrency mining produced more emissions than many industrialized nations. The contradiction was not a bug in implementation. It was a feature of the underlying physics.

Electronic transistors generate heat when they switch. More computation means more switching means more heat means more cooling means more power. Proof-of-work security requires computation that is deliberately wasteful—that is the point, the mechanism by which the network achieves consensus without central authority. But wasteful electronic computation is environmentally ruinous at scale. The mathematics of decentralized trust collided with the thermodynamics of electronic switching, and thermodynamics won.

The Light-Switch Ledger™—a blockchain architecture built entirely on photonic infrastructure—eliminates this contradiction entirely.

The numbers are stark.

A single Bitcoin transaction today consumes approximately 500 kilowatt-hours of electricity and generates roughly 500 grams of carbon dioxide. A single transaction on photonic infrastructure consumes less than 0.001 kilowatt-hours and generates less than 0.1 grams of carbon dioxide—less than a single human breath. The reduction is three orders of magnitude, achieved immediately, as a direct consequence of the underlying physics. But efficiency, however dramatic, is only the beginning. Speed transforms what blockchain can do.

Current blockchain architecture reflects the limitations of electronic computation. Bitcoin's ten-minute block interval exists because the network needs time to propagate transactions and reach consensus across thousands of nodes running electronic processors. The network must wait for information to travel around the world, for nodes to validate transactions, for miners to compete in the computational lottery that determines who adds the next block. These latencies make blockchain unsuitable for applications requiring real-time settlement.
Stock trades execute in microseconds; blockchain cannot compete. Payment authorizations must complete before a customer leaves the counter; blockchain cannot deliver. Micropayments could enable entirely new economic models—paying fractions of a cent for individual articles, songs, or API calls—but the transaction costs and confirmation times of current blockchain make such payments economically irrational.

Our Transaction Verification patent, filed January 2026 with 100 claims, addresses this directly. Using the Poovey Stack's validated switching speeds, financial transactions can be verified in approximately 65 nanoseconds—compared to 50-500 microseconds for electronic verification. ***This speed enables micropayments at $0.001 or less***, where current electronic fees of $0.02-0.30 make such transactions impossible. Block intervals can shrink from ten minutes to milliseconds. Throughput can scale from millions of transactions per second to billions. Our **Hash Engine** patent covers the photonic cryptographic computation that makes this possible—wavelength-multiplexed parallel processing that performs SHA-3 hashing at speeds electronic systems cannot approach. The combination creates blockchain infrastructure where security emerges from physics rather than energy waste, where trust is measured in femtoseconds rather than megawatts.

Picture the skyline of Atlanta, New York, Los Angeles, New Orleans, or Washington at night. Not just glass and steel—but photonic towers glowing with light. Inside them, trust and compute run silently at femtosecond speed. No cooling towers sending plumes of steam into the urban sky. No smokestacks marking industrial processes. No water lines draining municipal supplies. Just photons racing through logic gates, securing value and intelligence at the heart of the city.

• • •

The Network Becomes the Computer

The internet was built for a world that no longer exists.

When the protocols that govern global communication were designed in the 1970s and 1980s, computation was scarce and expensive while bandwidth was the bottleneck. The engineers who designed TCP/IP made rational choices given these constraints: concentrate computation in central facilities where expensive machines can be shared, use networks to move data to those facilities for processing, design the network itself to be simple and stateless—a dumb pipe that forwards packets without understanding or operating on their contents.

This architecture made sense when the limiting factor was processing power. It makes no sense when computation is cheap and bandwidth is precious, when the constraint is not the cost of processing but the time and energy required to move data across distances.

TCP/IP-L—our patent for photonic networking filed December 2025—implements computation within the network itself. Because the Poovey Stack processes information using light, and because optical networks already transmit information using light, the boundary between computation and communication dissolves. Network nodes become computational resources rather than passive relay points. The routers that forward your traffic can simultaneously process it. A signal traveling between two points need not remain unchanged during transit. Intermediate nodes can perform processing operations on the optical signal itself—filtering, pattern matching, preliminary inference, encryption, compression, transformation—without converting the signal from optical to electrical form and back again. The conversion between light and electricity is where electronic networks waste time and energy; optical-electrical-optical conversion at each network hop adds latency and consumes power. Photonic networks eliminate this conversion entirely.

The latency implications enable applications that current infrastructure cannot support. Consider remote surgery. A surgeon in Boston operating a robotic system on a patient in rural Montana needs visual feedback that accurately represents the current state of the surgical field. If the video feed is 60 milliseconds old, the surgeon's hands are operating on anatomy that has moved since the image was captured. Current telesurgery systems work only over short distances; photonic networks could make expert surgery available anywhere.

Consider autonomous vehicles. A self-driving car processing sensor data in a remote data center faces latency that translates directly into stopping distance. A photonic mesh that delivers single-digit millisecond latency nationwide changes the calculus of what autonomous systems can achieve. The infrastructure for this network can leverage existing assets.

The United States has approximately 350,000 billboard structures positioned along highways and arterial roads—the same corridors where fiber-optic cables run, where the physical infrastructure of the internet concentrates. These structures already have power connections and often fiber connectivity. Our patent portfolio includes technology for photonic repeaters scaled to fit such existing structures—devices roughly the size of a cable box that regenerate optical signals while performing computational operations on data in transit. Because these devices attach to unused structural real estate without interrupting the billboard's advertising function, the economics could prove far more attractive than building dedicated network infrastructure from scratch. A photonic repeater on an existing billboard requires no new construction, no years of permitting, no community opposition. The structure already exists with appropriate zoning. Deployment that would take a decade through conventional infrastructure development could proceed in a fraction of that time.

Combined with SkyLight towers serving as regional computational hubs, this architecture creates a network where computation and communication blur together, where data processing happens throughout the network rather than only at designated facilities. **The network becomes the computer.** *The dumb pipe becomes an intelligent fabric.*

• • •

Sovereign Infrastructure

The phrase "Golden Dome" has entered public discussion as shorthand for integrated air and missile defense—the systems that must detect, track, discriminate, and intercept threats ranging from legacy ballistic missiles to hypersonic glide vehicles to autonomous drone swarms. The technical requirements of these systems have escalated beyond what electronic infrastructure can deliver, creating capability gaps that adversaries are actively working to exploit.

The core challenge is sensor fusion. Modern air and missile defense does not rely on any single sensor. Radar systems detect objects at great distances but struggle to characterize them precisely. Infrared sensors can identify heat signatures but have limited range. Space-based sensors provide broad coverage but face limitations of resolution. Electronic intelligence intercepts communications but requires correlation with other data. Each sensor type provides partial information; effective defense requires combining these partial views into a coherent operational picture.

The quality of that combined picture depends critically on timing. Sensor data must be synchronized with precision measured in nanoseconds. A radar return from an object must be correlated with an infrared detection of the same object, and that correlation depends on knowing exactly when each sensor made its observation. If timing synchronization drifts by microseconds, correlation fails. Objects appear to duplicate; real targets merge with false ones; the operational picture degrades from clarity into confusion.

Photonic systems measure time in femtoseconds—a thousand times finer resolution than nanoseconds. This precision translates directly into sensor fusion capability. Track correlation becomes cleaner because timing windows become tighter. Spoofed sensor feeds—false data injected by adversaries to confuse the picture—fail time-and-spectrum checks that electronic systems cannot perform with sufficient precision.

Let me describe what the bleeding edge looks like—what becomes possible when validated femtosecond switching enters defense applications.

Hypersonic discrimination. A hypersonic glide vehicle traveling at Mach 20 covers roughly 7 kilometers per second. At electronic processing speeds measured in nanoseconds, the vehicle moves measurable distances between sensor samples—creating gaps in the track that adversaries can exploit with decoys and maneuvering. At femtosecond processing speeds, a thousand samples can be taken in the time the vehicle moves a single meter. The track becomes continuous rather than sampled. Maneuver signatures become readable. The behavioral patterns that distinguish warhead from decoy become visible in data that electronic systems blur into ambiguity.

Swarm defeat. Autonomous drone swarms present a different challenge: not a single fast-moving target but hundreds or thousands of slower targets acting in coordination. Electronic systems face a combinatorial explosion—the number of possible threat correlations grows factorially with swarm size, overwhelming processing capacity. Photonic parallelism, using the wavelength-division multiplexing architecture protected in our **Hash Engine and Outcome-Verifiable Platform patents**, can process sixty-four or more independent tracking channels simultaneously through a shared optical substrate. The swarm's coordination becomes visible; the command structure reveals itself in timing correlations that electronic systems lack the bandwidth to detect.

Electronic warfare at optical speed. Current electronic warfare systems adapt slowly—changing jamming frequencies or filtering profiles in response to observed threats over timescales of milliseconds to seconds. Adversaries exploit this by frequency-hopping faster than defensive systems can adapt. Photonic systems can reconfigure in femtoseconds. A frequency-hopping emitter that escapes electronic countermeasures becomes trackable when the defensive system can adapt a thousand times faster than the emitter can hop. The physics reverses; the advantage shifts from attacker to defender.

Cyber-physical security. The command-and-control systems that coordinate defensive responses face continuous cyber assault. Electronic C2 systems rely on cryptographic security that sophisticated nation-state attackers can potentially compromise. Photonic infrastructure provides security properties that electronic systems cannot match. Hardware-bound identity ensures that devices are what they claim to be—the optical properties of a photonic module constitute an unforgeable fingerprint. Continuous attestation verifies that systems remain uncompromised throughout operation. Our Outcome-Verifiable Platform patent covers the cryptographic attestation architecture that binds computational outcomes to auditable records. Any physical tampering changes measurable optical properties and triggers automatic response.

Post-quantum communications. The threat of quantum computing has loomed over cryptographic security for years. Sufficiently powerful quantum machines could break the mathematical foundations of current encryption. Our Hash Engine patent addresses this not through post-quantum mathematical algorithms but through the physics of photonic switching itself. The hash function exploits properties of light that quantum computers cannot accelerate—interference patterns, polarization states, and temporal correlations that have no mathematical shortcut. Security becomes a property of physics rather than a bet on computational difficulty.

The SkyLight National Photonic Mesh provides these capabilities as a ground layer for national defense infrastructure—faster to deploy than space-based assets, harder to spoof than electronic systems, easier to integrate into existing command structures than entirely new architectures. A forty-story tower in downtown Atlanta or a converted hotel in Washington can host SCIF-grade photonic floors close to decision-makers, with sovereign custody of keys, compilers, and audit trails maintained entirely under U.S. control.

• • •

Unlike current chip manufacturing, photonic computing does not require the concentrated fabrication capability in Taiwan and South Korea that makes today's semiconductor supply chain a geopolitical vulnerability. Our Relaxed Lithography patent means that photonic devices can be manufactured using mature processes available at dozens of facilities in the United States, Europe, Japan, and other allied nations. The strategic advantage, if achieved, depends on technology and facilities that can be built, maintained, and expanded on sovereign territory—immune to supply chain disruptions, blockades, or conflicts that could paralyze electronic alternatives.

I am not a defense contractor, and True Photonic is not positioning itself to become one. But the capabilities that emerge from photonic computing have obvious implications for national security. The companies that dominate defense technology—Lockheed Martin, Raytheon, Northrop Grumman, General Dynamics—have the security clearances, the procurement relationships, and the program management expertise that a technology startup cannot replicate. Our role is to provide the underlying platform—the physics, the manufacturing relationships, the operational expertise—that makes new capabilities possible.

The broader strategic point extends beyond any specific weapon system. A nation that controls femtosecond-class computational infrastructure—that can process information faster than any adversary, that can fuse sensor data with precision adversaries cannot match, that can secure communications through physics rather than just mathematics—occupies a position of technological advantage comparable to historical advantages in other domains. Air superiority transformed warfare in the twentieth century. Information superiority through photonic infrastructure could prove equally decisive in the twenty-first.

This strategic position enables what might be called cognition-as-export: a model where the United States and its allies control the most advanced computational infrastructure, and the outputs of that infrastructure—the intelligence, the analysis, the capabilities—reach global markets while the underlying technology remains sovereign - *A true America First technology strategy.*

The parallel to other strategic industries is instructive. The United States does not export its most advanced fighter aircraft, but it does export the security that those aircraft provide to allies. Pharmaceutical companies do not export their manufacturing processes, but they do export the drugs those processes produce. The value flows outward; the capability remains controlled.

Photonic computing infrastructure could operate on similar principles. A query submitted from anywhere in the world could be processed on sovereign infrastructure and the result returned—the intelligence exported, the means of production retained. Nations that control this infrastructure do not need to restrict access to AI capabilities; they can provide access on terms that reflect who owns the underlying platform. This is not isolation but competitive advantage of the kind nations have historically sought to maintain in strategically significant industries. The implications extend to AI development itself. If the most capable AI systems require photonic infrastructure to train and operate efficiently, then control of photonic computing becomes control of the AI frontier. The nation or alliance that masters this technology does not merely gain military advantage; it gains the ability to shape how artificial intelligence develops globally. The models trained on sovereign infrastructure, the capabilities those models enable, the applications built on those capabilities—all flow from control of the computational foundation.

• • •

Devices That Live Inside Us

The applications described so far operate at the scale of nations and networks. But the same physics that enables planetary-scale transformation also enables interventions at the most intimate scale imaginable: inside the human body. The heart generates electrical signals that can be read from the skin's surface. Modern artificial intelligence can detect patterns in cardiac electrical activity that human cardiologists cannot perceive—subtle variations in waveform morphology, timing relationships between beats that emerge only in statistical analysis, trends that develop over hours or days and predict events that have not yet occurred. The limiting factor is no longer algorithmic capability. The limiting factor is the hardware that runs these algorithms. Continuous AI analysis requires continuous computation. Continuous computation in electronic devices requires continuous power consumption and generates continuous heat.

Current wearable cardiac monitors face a fundamental constraint. A processor running sophisticated neural networks continuously will exhaust its battery in hours. Manufacturers compromise by running simplified algorithms—rule-based detectors that miss subtle patterns—or by running sophisticated algorithms intermittently, missing transient events that occur between analysis windows. CardioLT—protected by our December 2024 patent filing—eliminates this compromise. A photonic processor generates negligible heat because photonic switching generates negligible heat. The processor can run sophisticated neural networks continuously without becoming warm, without degrading battery life through heat-induced inefficiency, without creating discomfort for the wearer. The same power efficiency that enables ninety percent reduction in data center energy consumption enables fourteen-day battery life in a wearable cardiac monitor measuring thirty-five by twenty by two millimeters. The patient wears a device smaller than a standard adhesive bandage. The device monitors every heartbeat, continuously, with AI sophistication previously available only in hospital intensive care units. When it detects concerning patterns, it transmits alerts and diagnostic data without requiring patient action. The patient need not notice symptoms; the device notices patterns the patient cannot feel. The regulatory pathway for CardioLT is 510(k) clearance—the FDA's process for devices substantially equivalent to existing approved products. Cardiac monitors exist and have been cleared through this pathway many times. CardioLT is a better cardiac monitor; the regulatory path is well-understood.

• • •

Inside the body, the constraints become more severe and the opportunities more profound.

Implantable cardiac devices—pacemakers that regulate heart rhythm, defibrillators that shock hearts out of dangerous arrhythmias—have saved millions of lives. They are also limited by constraints that have persisted for decades. Battery life determines how often patients must undergo surgery to replace depleted devices. Current pacemakers last seven to ten years; defibrillators last five to seven years. A patient implanted at age sixty will face three or four replacement surgeries over a normal lifespan—each carrying risks of infection, bleeding, and complications. Heat generation limits computational sophistication. An implanted device that becomes even slightly warm damages surrounding tissue. Current devices therefore run the simplest algorithms that can achieve acceptable performance—algorithms that have not fundamentally changed in design philosophy for decades. The result: ten to fifteen percent of defibrillator discharges are triggered by rhythms that were not actually dangerous. The device mistakes a rapid but benign rhythm for a lethal arrhythmia and delivers a shock. These shocks are painful—patients describe them as being kicked in the chest by a horse. Some patients request removal despite the life-threatening conditions the devices were implanted to treat. They would rather risk sudden cardiac death than continue living with a device they cannot trust.

PulseLT—protected by our December 2024 patent filing—addresses each of these constraints through photonic architecture. A processor generating effectively zero heat can run sophisticated AI continuously without tissue damage. Deep learning models trained on millions of cardiac rhythm examples can distinguish dangerous ventricular arrhythmias from benign rapid rhythms with accuracy far exceeding simple threshold comparisons. Inappropriate shocks could be reduced by ninety percent or more. Power consumption one-hundredth of electronic equivalents enables battery life exceeding twenty-five years—potentially the patient's lifetime. A patient implanted at age sixty-five might never need replacement surgery. The device lasts as long as the patient does.

Optical fiber leads—thin, flexible glass fibers that carry light rather than electricity—eliminate the MRI heating problem that restricts three-quarters of cardiac device patients from needed imaging. Light does not interact with magnetic fields the way electrical current does. A patient with optical leads can undergo MRI scanning without restriction. The regulatory pathway for PulseLT is more demanding. Implantable cardiac devices are Class III medical devices requiring Premarket Approval—the FDA's most rigorous pathway, involving clinical trials demonstrating safety and efficacy in human patients. The timeline is measured in years, not months. But the destination, if reached, transforms a major category of medical technology and eliminates suffering that current devices cause while trying to prevent it.

• • •

The Photonic Shape-Shifting Muscle

The Photonic Muscle patent—filed November 2024—extends the platform beyond sensing and processing to actuation: **making things move and making them faster and stronger.**

The technology uses light-activated piezoelectric elements to achieve mechanical response at femtosecond timescales—a million times faster than the fastest electronic motor controllers. This speed might seem academic; no mechanical system needs to respond in femtoseconds. But the speed of the controller translates into control bandwidth—how rapidly a system can sense, decide, and correct. A robotic arm controlled by electronic processors can update its position perhaps a thousand times per second. Between updates, the arm moves open-loop, without feedback, accumulating errors that the next update must correct. A photonically controlled arm can update a million times per second. Errors have no time to accumulate. Corrections happen before deviations become significant. *Consider a surgical robot positioning a scalpel.* The surgeon's hand trembles slightly—an unavoidable consequence of human physiology. Electronic control systems can filter some of this tremor, but high-frequency components persist between samples. Photonic control, operating a thousand times faster, can cancel tremor more completely, enabling steadier cuts, smaller incisions, access to anatomy too delicate for current robotic surgery.

Consider a prosthetic hand grasping an egg. Current electronic prosthetics achieve stable grasping through careful calibration and slow, deliberate motion. Users learn to move carefully because the hand cannot respond quickly enough to dynamic changes. Photonic prosthetics could achieve stable grasping dynamically—responding to incipient slip before the egg begins to fall, adjusting grip force preemptively rather than reactively.

Consider an exoskeleton augmenting a worker's strength. Current exoskeletons add force but also add lag—a perceptible delay between the user's intention and the machine's response. Users report fighting their exoskeletons rather than being helped by them. Photonic controllers, operating in nanoseconds rather than milliseconds, could respond faster than human nerves transmit signals. The exoskeleton could predict and assist motion before the user's own muscles fully engage. The insight that unifies these applications is that photonic muscles require photonic brains. An actuator capable of femtosecond response, controlled by a processor with microsecond response, wastes most of its capability. The Poovey Stack provides the control system that allows photonic actuators to achieve their theoretical performance. The combination creates robotic and prosthetic systems operating in regimes that electronic systems cannot reach.

• • •

The Ecosystem

We established **True Photonic Ventures** to pursue these types of verticals. *Not all of them—no single company could pursue them all*—but enough to demonstrate that photonic computing is not merely more efficient infrastructure. It is a new category of capability, and we intend to own as much of that category as the patent system allows. The structure creates and supports vertical-specific companies. The parent company provides technology licenses, technical support, and access to the photonic platform. The ventures arm provides capital, business development expertise, and the infrastructure for building companies in specific markets. The resulting portfolio companies combine photonic capability with domain expertise, pursuing opportunities that neither a pure technology company nor a pure domain company could capture alone.

The physical infrastructure strategy creates additional alignment. Portfolio companies become tenants in SkyLight facilities, locating their computational workloads in Clean Compute Centers built around photonic infrastructure. CardioLT's AI training runs on SkyLight servers. The Light-Switch Ledger validates on SkyLight infrastructure. The ecosystem reinforces itself: more applications create demand for more infrastructure; more infrastructure enables more applications.

Beyond the filings already made, we are actively preparing applications that extend the Poovey Stack into additional high-value verticals—from real-time graphics rendering and autonomous vehicle decision systems to next-generation data center architectures and the eventual bridge to quantum computing.

These filings will complete a portfolio covering every major frontier where computing performance has become the binding constraint.

• • •

The applications in this chapter span from the planetary to the personal. A new architecture for money and trust that eliminates the environmental catastrophe of current blockchain. A new model for the internet that transforms passive transport into active computation. Sovereign infrastructure for national defense that provides capabilities electronic systems cannot match. Medical devices that operate inside the body for decades rather than years. Robotic systems that respond faster than human nerves can transmit signals.

They are unified by the platform that enables them: the Poovey Stack's combination of femtosecond speed, negligible heat generation, extreme power efficiency, and operation in environments hostile to electronics. Each application exploits these characteristics in different ways, but all depend on the same underlying physics. The same switch that enables a sustainable blockchain enables a twenty-five-year pacemaker. The same processing capability that hardens missile defense enables surgery more precise than human hands can perform.

The honest acknowledgment that has characterized this book applies to these applications as to the infrastructure chapters. The Light-Switch Ledger is in development, not deployment. TCP/IP-L is a patent filing, not a shipping product. The defense applications require partnerships and clearances we do not yet possess. CardioLT and PulseLT are pursuing regulatory pathways, not holding FDA clearances. Photonic Muscle is a patent portfolio, not a product line. Each application requires years of engineering, regulatory navigation, market development before it can deliver on its potential.

But the potential is grounded in physics that has been validated, not speculation about undiscovered phenomena. The femtosecond switching speeds that enable the Light-Switch Ledger have been measured at Technion and in True Photonic's laboratories. The power efficiency that enables twenty-five-year implant battery life has been demonstrated at the device level. The applications follow from the technology; the question is execution, not feasibility.

True Photonic Ventures exists to pursue that execution across multiple verticals simultaneously. Some ventures will succeed; others may not. But the platform that underlies them all—the Poovey Stack, the SkyLight infrastructure, the photonic computing capability—provides a foundation for continued innovation regardless of which specific applications reach market first.

The next generation of everything.

Not hyperbole.

Aspiration.

The final chapter examines the path from current status to this envisioned future: the timeline, the milestones, the obstacles that remain, ***and the reasons for believing that 2026 may indeed become the year of light.***

CHAPTER 14

The Light Ahead

The moment that set everything in motion was not a breakthrough in the laboratory. It was a realization about a person.

Dr. Gary Poovey had spent twenty-five years working with leading semiconductor manufacturers, developing the plasma physics and materials science that would eventually produce the light-actuated light-switch at the heart of photonic computing. He had solved the problem that defeated every previous attempt at optical logic—how to make light control light at speeds that matter. And I realized, watching him work, that he might fail anyway. Not because the physics was wrong, but because the path from laboratory discovery to world-changing technology requires more than physics. It requires capital, organization, strategy, protection, and the willingness to fight battles that scientists are not trained to fight.

Gary Poovey is a wonderful man. He is also a man whose work might never have been known if someone had not recognized what he had accomplished and committed to ensuring the world would see it. That someone turned out to be me—not because I understood the physics better than anyone else, but because I understood what happens to important work when no one protects it. It gets buried. It gets forgotten. It gets acquired by larger players who sit on it to protect their existing businesses. The history of technology is littered with breakthroughs that arrived too early, or too quietly, or without the organizational muscle to survive the journey from laboratory to market.

I was not going to let that happen to photonic computing.

• • •

The previous chapters have described a technology that exists, at least in validated prototype form, and that has continued to advance since initial testing. The fundamental switching physics were validated at Technion University in Israel, where pump-probe spectroscopy measured switching speeds in the 150-200 femtosecond range—three to four times faster than anything previously published in the scientific literature, and five thousand times faster than the electronic transistors that power contemporary computing. Since that validation, the technology has continued to evolve. The original switch composition has been reformulated to eliminate lead, addressing both environmental concerns and regulatory requirements for commercial deployment. The graphene-boron nitride-sapphire substrate represents a further advancement, offering thermal conductivity exceeding 2,000 watts per meter-kelvin and enabling the dense three-dimensional architectures the Poovey Stack requires. The Relaxed Lithography approach offers a manufacturing pathway that bypasses the most constrained chokepoints in the global semiconductor supply chain. The question that remains is not whether the underlying physics works in a laboratory setting, but whether the complete system can be brought to market, at what pace, and against what obstacles.

The intellectual property position now includes fifteen patent filings with over 800 claims spanning the core switching technology, manufacturing processes, substrate development, memory architecture, network infrastructure, AI inference, cryptographic computation, transaction verification, and multiple vertical applications including cardiac monitoring, implantable devices, and robotic actuation. This portfolio builds on published prior art including the NIBLE BUS—an all-optical architecture for managing data transfer through photonic switches—and a purely optical NAND logic gate that establishes the foundational building block from which all Boolean operations can be constructed. The patents filed between November 2024 and January 2026 represent a systematic effort to own the intellectual framework that will define photonic computing before the market understands what is coming.

The organizational structure reflects a parallel development strategy. True Photonic Inc. develops the core technology platform. SkyLight Holdings develops the infrastructure to deploy that technology—identifying and converting stranded tower assets, securing power and connectivity, preparing facilities to receive photonic processing systems. True Photonic Ventures develops vertical applications that demonstrate specific use cases. Each entity has its own timeline, its own milestones, and its own path to revenue—but all three advance concurrently, with progress in each reinforcing the others.

When the machines are ready, ***the data centers must be ready to receive them***. When the platform matures, the vertical applications ***must be prepared to utilize it.***

What We're Up Against

The obstacles that remain are substantial, and intellectual honesty requires acknowledging them directly rather than minimizing their significance. Some obstacles are technical, some commercial, some regulatory, and some relate to the inherent difficulty of displacing entrenched technologies regardless of the new technology's merits.

The technical obstacles begin with integration complexity. Individual photonic switches have been validated, and the physics of combining them into logic gates is well understood. But building complete processors—with billions of interconnected elements, clock distribution, power management, and interface circuitry—involves engineering challenges that become apparent only at scale. The transition from working devices to working systems has defeated promising technologies before. Photonic computing is not immune to this risk. Manufacturing yield presents another challenge. The Relaxed Lithography approach reduces yield risk by using mature processes with well-characterized defect densities, but photonic devices impose requirements that differ from electronic transistors in ways that could introduce unexpected yield loss mechanisms.

The commercial obstacles may prove more challenging than the technical ones. Established semiconductor companies—Intel, AMD, NVIDIA, and others—have massive resources, existing customer relationships, and strong motivation to defend their market positions. They could attempt to acquire photonic computing technology to control its development pace, develop competing approaches that leverage their existing manufacturing assets, or simply compete on price by accepting lower margins to delay photonic adoption.

NVIDIA in particular presents a formidable competitive position. The company's CUDA ecosystem has created software lock-in that makes switching architectures costly for customers even when alternative hardware offers theoretical advantages. NVIDIA GPUs are not merely products but platforms, with thousands of engineers trained on their specific programming models and millions of lines of code optimized for their specific capabilities. Any alternative computing approach must either integrate with this ecosystem or offer advantages so compelling that customers willingly accept the switching costs. Photonic computing's performance advantages may eventually reach that threshold, but the path from laboratory validation to ecosystem displacement is neither short nor certain. Customer adoption presents its own obstacles. Even if photonic processors deliver the performance and efficiency advantages the physics predicts, customers must modify their software, retrain their engineers, and accept the risk of deploying unproven technology in production environments. Enterprise technology decisions involve committees, procurement processes, and risk assessments that favor proven vendors and established approaches.

The response to adoption risk is not to wait for customers but to become one. TrueGil™, our gaming subsidiary, addresses this directly. Gaming demands exactly what photonic computing delivers: extreme speed, low latency, high throughput. By building our own use case, we create a development platform that proves the technology in production environments we control—while generating revenue independent of enterprise sales cycles. We do not need to convince skeptical procurement committees to take a chance on unproven technology. We prove it ourselves, in public, with real users. The performance data from TrueGil™ becomes the case study that enterprise customers require.

Regulatory and certification challenges affect different parts of the business differently. Medical devices like CardioLT and PulseLT require FDA clearance processes that add years to development timelines. Data center facilities require permitting that varies by jurisdiction and has become increasingly contentious. Network infrastructure deployment involves the FCC, state public utility commissions, and local zoning authorities. None of these obstacles is insurmountable, but together they constrain the pace at which even successfully developed technology can reach the market.

The Skeptics

The most persistent skepticism comes not from technologists who doubt the physics but from the investment community, whose institutional caution serves a purpose but also creates a structural blind spot. The pattern is familiar to anyone who has raised capital for deep technology.

Investors want proof before they commit. They want validated prototypes, then pilot customers, then revenue, then profitability—each milestone unlocking the next conversation, each round of funding contingent on demonstrating that the previous round's targets were met. This discipline protects capital from speculative ventures that cannot deliver.

It also systematically disadvantages technologies that require patient development before they can demonstrate commercial metrics. The venture capital industry has shifted toward software investments with shorter cycles and more predictable returns. Deep technology—the kind that requires building physical things, navigating regulatory approval, and waiting for manufacturing processes to mature—fits poorly into fund structures designed for quick deployment and quicker returns. The investors who specialize in hard technology understand semiconductor development timelines, but they are fewer than in previous generations, and competition for their attention is intense.

The skeptic's argument is reasonable on its face: wait for more proof, reduce risk, invest when the technology is further along. But this argument contains a timing fallacy. By the time photonic computing has demonstrated everything the skeptics require, *the strategic window will have closed.* The intellectual property will have been filed. The manufacturing relationships will have been established. The first-mover advantages will have been captured. The investors who waited for certainty will find that certainty arrived too late to matter.

My answer to the skeptics is not to argue but to execute. While they wait for proof, we file patents. While they demand more validation, we secure manufacturing partnerships. While they deliberate, we build. The fifteen provisional patents filed over the past fourteen months represent not just intellectual property protection but a strategic response to institutional caution: if the investment community will not move at the pace the opportunity requires, we will own enough of the future that their eventual participation will be on terms that reflect who took the risk when it mattered. And while they wait for enterprise customers to validate the technology, TrueGil™ becomes our own customer—proving performance in production, generating revenue, and building the case study skeptics require.

Why Now

Against these obstacles, why suggest that 2026 could become the year of light? The proposition rests not on technological optimism alone but on the specific convergence of circumstances that the computing industry now faces.

The infrastructure crisis described in Part I of this book has reached an inflection point. Power constraints have become the limiting factor for AI development at every major technology company. Microsoft's Stargate project requires nuclear power plants. Google data centers draw more electricity than some countries. Meta, Amazon, and OpenAI have all announced infrastructure investments measured in hundreds of billions of dollars, yet executives at each company acknowledge that they cannot build fast enough to meet projected demand. This is not a theoretical future problem; it is a constraint affecting product decisions today. Any technology that can deliver comparable computational throughput at dramatically lower power consumption commands immediate strategic interest regardless of its maturity level.

The scale of unmet demand creates market opportunity that historical semiconductor transitions never enjoyed. When Intel introduced the microprocessor, the personal computer market did not yet exist; the company had to create demand for its new capability. Today's AI infrastructure market is neither nascent nor uncertain—it is constrained by supply rather than demand. Companies are paying premium prices for computing capacity, signing multi-year commitments for hardware that does not yet exist, and still facing allocation limits. Into this environment, a technology that offers ten times the throughput per watt does not need to create demand; it needs only to demonstrate that it works.

The supply chain vulnerabilities have become impossible to ignore. TSMC's concentration in Taiwan represents a geopolitical risk that corporate boards and government agencies now discuss explicitly. The ASML chokepoint creates bottlenecks that no amount of spending can quickly resolve. An alternative approach that sidesteps these constraints—that can manufacture using widely available process nodes rather than contested leading-edge capacity—addresses strategic concerns that have escalated dramatically in recent years.

The community backlash against data centers has transformed the regulatory environment. Projects that would have sailed through permitting five years ago now face organized opposition, extended review processes, and outright rejection. More than one hundred billion dollars in planned data center investment has been blocked, delayed, or significantly modified due to community concerns about water consumption, energy usage, and resource extraction.

A technology that can deliver data center capability without the resource footprint that generates opposition changes the calculus. Communities that reject electronic data centers might welcome photonic facilities that require no cooling towers, minimal water consumption, and a fraction of the grid capacity.

• • •

The Alternative to Success

The honest assessment of risk requires considering not just whether photonic computing might fail, but what happens if no one solves the problem it addresses.

The alternative is not the status quo. The alternative is not electronics continuing to improve along familiar curves. *Those curves are bending.* Moore's Law has not died, but it has slowed, and the economic engine that drove it—smaller transistors enabling both better performance and lower cost—has broken. Leading-edge chips now cost more per transistor than their predecessors, not less. The physics of electron-based switching is approaching limits that cannot be engineered around. Heat dissipation scales with transistor density in ways that cooling technology cannot indefinitely address.

If photonic computing does not work—if this approach and every other alternative fails to extend computational capability beyond what electronics can deliver—*the consequence is contraction.* ***AI development slows.*** The capabilities that the past decade has demonstrated plateau rather than expand. The promise of artificial general intelligence, of scientific breakthroughs accelerated by machine learning, of problems solved by computational approaches that do not yet exist—these possibilities recede rather than approach.

This is not fearmongering.

It is the logical consequence of physical constraints that cannot be wished away. The computing industry has operated for sixty years on the assumption that next year's hardware will be better than this year's. Business models, research agendas, and national strategies all embed this assumption. If the assumption fails—if computational progress stalls because the underlying technology has reached its limits—the consequences propagate through every domain that depends on continued improvement. Photonic computing is not the only possible answer. Quantum computing, neuromorphic architectures, and approaches not yet imagined might also extend the trajectory.

But photonic computing is the approach that has demonstrated validated switching speeds, that has a manufacturing pathway through existing infrastructure, that addresses the specific constraints—power, heat, speed—that currently limit progress. If it works, it works soon enough to matter. That is not a guarantee of success. ***It is a reason to try.***

• • •

What I Have Bet

The abstract discussion of risk becomes concrete when I consider what I have personally committed to this effort.

This work has strained relationships with my children, who have watched their father pursue something they cannot fully evaluate and cannot easily support. They have seen this pattern before. Derek Automotive was working—gaining traction, building momentum—and I shifted focus to something bigger. Staffing was working—profitable, growing, reliable—and I redirected that success toward another challenge. From their perspective, the question is fair: *when do we come first?* When does the next opportunity stop being more important than the stability we thought we had built? I do not have an answer that would satisfy them, because the honest answer is that I cannot look at something like photonic computing—something that could matter at civilizational scale—and choose smaller ambitions because they are safer.

That is who I am. *It is not always who they need me to be.* It has stretched the patience of my Derek Automotive investors, who committed capital to one venture and have watched my attention divide toward another. It has absorbed all my capital—**not some of it, all of it.**

I have asked gifted professionals to work on the come, accumulating obligations I will be personally responsible for if this does not work out. The relationships at stake are not merely professional; some could be damaged beyond repair if I fail to deliver on the trust these people have placed in me.

I am not a young founder with decades to recover from failure. The trust placed in me to accomplish something this significant means that failure does not merely fail a business plan—it fails that trust in ways that will define the final chapter of my professional life.

I will go out a winner or a loser.
That is the temperature of my last act.

I say this not to elicit sympathy but to establish stakes. When I argue that this technology can work, that this timeline is achievable, that these obstacles can be overcome, I am not making claims I can walk away from if they prove wrong. My credibility, my relationships, my financial security, and my legacy are all committed. ***That commitment does not make me right.*** But it should make clear that I am not speculating casually about possibilities I will not personally experience.

New Orleans holds particular significance in this commitment. The city of my birth, where family still lives, where the people who will work in a SkyLight facility include people I have known my entire life. I cannot treat New Orleans as an abstraction on a spreadsheet. If photonic computing can deliver what the physics promises, I want that delivery to begin in a place where I will personally witness the consequences—**good or bad.** The aspiration is to make New Orleans the first city to deploy operational photonic computing infrastructure. The city that recovery has rebuilt repeatedly could become the city that demonstrates a new model for computational infrastructure: one that creates jobs, preserves buildings, and operates without the resource burdens that have made data centers unwelcome elsewhere.

June 2027
Predictions without specifics are not predictions; *they are hopes.*

Here is what I am committing to:
By June of 2027, we will be in a urban tower, doing logic-in-light.

That means a physical facility—a converted building of the kind described in Chapter 12—operational and running photonic processors that perform real computation.

Not a laboratory demonstration.

Not a prototype in a controlled environment.

A working system in a commercial facility, processing actual workloads, demonstrating that the technology validated at Technion can be manufactured, deployed, and operated at the scale that matters.

This benchmark is aggressive. It requires that device fabrication succeed, that system integration work, that facility conversion complete on schedule, that the thousand things that can go wrong in technology development either do not go wrong or get solved quickly when they do. The people working on this effort believe it is achievable; they would not have committed their time and reputations otherwise. But belief and achievement are different things, and only June 2027 will reveal which we have.

I am stating this benchmark publicly because accountability matters. If we achieve it, the skeptics will have their answer. If we do not, I will have to explain why—to investors, to partners, to the readers of this book, and to myself. That exposure is intentional.

The future belongs to those willing to be wrong in public.

• • •

The vision that emerges from this analysis is neither triumphalist nor modest. It suggests that the computing industry stands at an inflection point comparable to the transitions from vacuum tubes to transistors, from mainframes to personal computers. Each of those transitions seemed unlikely to many observers even as it was occurring. Established players had reasons to believe their positions were secure. New approaches faced technical obstacles that skeptics considered insurmountable. The transitions happened anyway, driven by fundamental economics that eventually overwhelmed institutional resistance.

The economics driving this potential transition are straightforward even if the technology is complex. A photonic processor that consumes one-tenth the power of an electronic processor to perform the same computation—if such a device can be built and manufactured at reasonable cost—eventually displaces the electronic processor regardless of installed base, existing relationships, or customer inertia. A data center that requires no cooling towers and minimal grid capacity can operate in locations and under conditions where electronic data centers cannot. A network that processes data in transit rather than merely transporting it enables applications that today's infrastructure cannot support.

The appropriate posture for those advancing this technology is humility rather than certainty. The gap between laboratory validation and commercial success has swallowed many promising innovations. The obstacles are real, the timeline is aggressive, and the competition includes some of the world's most capable and well-resourced technology companies. Nothing in this book should be taken as a guarantee that photonic computing will succeed, that True Photonic specifically will succeed, or that the timeline suggested here will prove accurate.

What can be said with confidence is that the current trajectory of computing infrastructure is unsustainable. Power consumption cannot continue increasing at current rates. Water usage cannot continue depleting aquifers and straining municipal systems. Community opposition cannot be indefinitely overridden. Supply chain vulnerabilities cannot be wished away. Something will change, either because better technology enables a different approach or because demand for computation contracts to fit what current technology can supply.

The proposition of this book is that light offers a better path than retreat. The physics of photons enables computational capabilities that electrons cannot match. The materials science has advanced to the point where practical devices can be fabricated. The manufacturing approach leverages available infrastructure rather than requiring constrained resources. The deployment model transforms stranded assets into productive facilities while resolving rather than creating community conflicts. The vertical applications address real needs with technology that could not exist in electronic form.

The title of this book—**Keep Computing**—contains both a **description** and an **aspiration**. It describes a technology that enables computational progress to continue beyond the limits of electronic approaches. It aspires to a future where the benefits of artificial intelligence, machine learning, and advanced computation remain available to those who seek them, without requiring environmental sacrifice, community disruption, or dependence on vulnerable supply chains.

The year 2026 into 2027 may or may not become the year of light. What happens next depends on execution, on resources, on factors beyond any single organization's control. But we know the **physics is sound**, the **core technology is validated**, and the need is **urgent**. *That is enough to justify the work ahead.*

The future belongs to those willing to be wrong in public.

I am willing.

They said photonic computing was fifty years away. They were measuring in the wrong units.

AFTERWORD

Infrastructure That Shares

In the summer of 2021, residents of Mesa, Arizona discovered that their water rates would increase by twenty-five percent over the next five years. The reason was infrastructure investment—not for their homes or businesses, but for data centers. Google, Apple, and Meta were building facilities in the region, drawn by cheap land and cheaper power. The aquifers that served the community would now serve the cloud.

Mesa's residents had no stake in the facilities draining their water. They would not share in the profits when those data centers processed queries, trained models, and ran the computations that powered trillion-dollar companies. They would simply pay more for water while the economic value flowed to shareholders in California.

This pattern is not new. It is the pattern of every extractive economy in history.

• • •

In the coal regions of Appalachia, communities hosted an industry that powered the nation. They absorbed the pollution, the mine collapses, the black lung disease. The coal left on trains and did not come back. The profits accumulated elsewhere. When the industry declined, it left behind poisoned streams, collapsed local economies, and populations too sick and too poor to leave. In the oil towns of Texas and Oklahoma, the pattern repeated. Roughnecks worked the rigs, their towns swelled with temporary prosperity, and the wealth flowed to Houston, to shareholders, to anywhere but the communities that hosted extraction. When the wells ran dry, the towns became shells.

The digital economy was supposed to be different. Information does not deplete. Data centers do not scar the land like strip mines. The technology industry promised a cleaner, more distributed prosperity. **That promise has not been kept.**

Today's data centers extract water from Arizona aquifers, electricity from Oregon dams, and land from rural communities across America.

The computation they perform generates trillions of dollars in value. That value accrues to the companies that own the infrastructure and to the shareholders who own those companies. The communities that host this infrastructure receive jobs, temporary construction booms, and modest tax revenues negotiated from positions of weakness. They do not receive ownership. They bear the costs and watch the benefits flow elsewhere.

This is the extractive model. **It is a choice,** not an inevitability.

• • •

There is a reason the model persists. Ownership requires capital, and communities do not have the capital to invest in semiconductor fabs or data centers. They cannot write billion-dollar checks. They can only offer what they have: water, power, land, and regulatory accommodation. They trade resources for jobs and hope the exchange is fair.

But what if the exchange could be structured differently?

Photonic Compute Credits represent access to compute capacity on our infrastructure. One credit equals one megawatt-hour of renewable-powered photonic compute capacity—real infrastructure, measurable and verifiable. Credits can be purchased, held, transferred, and redeemed. This is not a cryptocurrency designed for speculation. It is a mechanism for distributing ownership of infrastructure access rights in a way that traditional equity structures cannot achieve.

A town of fifty thousand people cannot collectively purchase shares in a private company. The legal structures do not exist. The minimum investment thresholds exclude them. But a town of fifty thousand people could receive credits representing access rights to compute infrastructure built in their community. A local foundation could hold credits and resell access to regional businesses, keeping the margin as ongoing revenue. A municipality could accept credits as partial compensation for hosting facilities and use them to subsidize compute access for local schools and hospitals.

These are not theoretical possibilities. **They are design choices.**

• • •

I want to be clear about what we are not claiming.

We are not claiming that this model solves inequality. The structural forces that concentrate wealth are larger than any single company's policies. Credit-based community ownership is a mechanism, not a revolution.

We are not claiming that we have all the answers. The model is still in development. The specific allocation percentages, governance structures, and eligibility criteria will be refined through experience. **We will make mistakes.**

We are not claiming purity of motive. True Photonic is a business. We need to generate returns for our investors, or we will not exist long enough to implement any model at all. The community ownership structure is designed to be compatible with investor returns, not to replace them.

What we are claiming is simpler: **the choice exists.** Technology companies can structure ownership to include the communities that host their infrastructure.

The tools exist. **What has been lacking is will.**

• • •

In the opening chapter of this book, I described **water that did not come back.** A community's allocation strained to serve a cloud it would never own. That image stayed with me as we built True Photonic—not because photonic computing eliminates water use entirely, but because it captured something essential about the relationship between infrastructure and community. The water belonged to the community. **The computation belonged to someone else.** *The benefits flowed away, like the water itself.*

I do not want True Photonic to be that kind of company.

In the Prologue, I described how Gary Poovey came to join us. What I did not fully explain was the condition he placed on that partnership. Gary prayed with his family before he committed. When he called back with his answer, it came with a request: he wanted to work with people who would build an ethical company. He had spent his career in an industry that optimized for shareholder returns above all else. He had watched good technology serve narrow interests.

He was entrusting us with his life's work—the materials science breakthrough that made everything in this book possible—and he wanted assurance that we would do some good in the world with it.

We made him that promise.

I am not a particularly virtuous person. I am a businessman who has spent forty years chasing opportunities wherever they appeared. But a promise is a promise. And beyond the obligation, I have built enough businesses to understand that extraction is not sustainable. Eventually, the community pushes back. Eventually, the political winds shift. Eventually, someone asks the question that should have been asked from the beginning: **Who benefits from all this?**

When that question is asked of photonic computing, I want the answer to be: ***everyone who helped make it possible.***

• • •

In the previous chapter, I described what I have personally bet on this effort. The strained relationships with my children, who have watched their father redirect one successful venture after another toward something they cannot fully evaluate. The absorbed capital. The professionals working on the come. The trust I will have betrayed if this does not work.

But I do not want to win alone.

If photonic computing succeeds, it will generate enormous value. That value should flow to the communities that host our facilities, the workers who build and operate them, the early believers who took risk when success was uncertain, the researchers who dedicated careers to problems that might never be solved. And yes—to the shareholders and founders who provide capital and bear risk. Their returns are legitimate. But they should not be the only returns.

The value photonic computing creates is too large, and the communities that make it possible are too important, for the old extractive model to be acceptable.

• • •

We are building something different. *We are building infrastructure that shares.*

Gary Poovey trusted us with his breakthrough because we promised to do good with it. The community ownership model is how we keep that promise.

Not through charity, not through foundations that bear our name, but through structures that give communities real power over real assets.

Ownership, not gratitude.

Stakes, not handouts.

The technology is ready. The ownership model is ready. What remains is execution—and communities willing to be partners rather than hosts.

We are looking for those communities now.

The future belongs to those willing to be wrong in public.
I am willing.

And I would rather fail trying to build something worth building than succeed at something that extracts more than it returns.

— **Derek W. Bailey**
January 2026

We built the cloud to escape the physical world.
The physical world is sending the bill.

I have bet everything I have on the physics being right. The physics does not care what I have bet.

A Note on Sources

The industry data, company valuations, and technical specifications cited in this book are drawn from public sources including company press releases, SEC filings, peer-reviewed scientific publications, and reputable financial databases. The Technion validation of the Poovey Switch is documented in Report #9105027635. Patent publications referenced are searchable in the USPTO database. Readers seeking additional detail on any specific claim are welcome to contact the author through True Photonic.

About the Author

Derek W. Bailey is CEO and Co-Founder of True Photonic, Inc., a photonic computing company developing femtosecond-scale optical switching technology independently validated at Technion University in Israel at 150-200 femtoseconds—three to four times faster than any previously reported result.

Bailey has been an entrepreneur for more than four decades, building ventures across hospitality, real estate, insurance, staffing, and technology. He has not received a paycheck that wasn't 100 percent commission since he was nineteen years old. His companies include OnPoint Staffing, Curbside Chef, International Broadband Technologies (video over copper wire), Popa Media (digital billboards with integrated VoIP), and WolvertonBailey, Inc., which developed the patented Counterpoise engine. He is also founder of Derek Automotive

Technologies, developing parallel-hybrid self-charging electric vehicles.
His twenty-seven-year partnership with Del Wolverton has produced multiple patents across ventures. When Dr. Gary Poovey entrusted them with his breakthrough materials science—the organic semiconductor architecture that makes photonic computing practical—Bailey structured True Photonic around a promise: to build an ethical company that shares value with the communities hosting its infrastructure.

True Photonic now leads a multi-entity structure including the core platform, SkyLight Holdings (Clean Compute Centers), and True Photonic Ventures (vertical applications including TrueGil™ gaming and medical devices). The patent portfolio spans fifteen filings with more than eight hundred claims.

Keep Computing is his first book.
Bailey was born in New Orleans and lives in Atlanta, Georgia.

Contact: dbailey@truephotonic.com

Made in the USA
Las Vegas, NV
10 February 2026

41641431R00111